ZEITSCHRIFT FÜR ALLGEMEINE MIKRO-BIOLOGIE

AN INTERNATIONAL JOURNAL ON MORPHOLOGY, PHYSIOLOGY, GENETICS, AND ECOLOGY OF MICROORGANISMS

HEFT 9 · 1982 · BAND 22

AKADEMIE-VERLAG · BERLIN

EVP 20,— M

ISSN 0044-2208 34112

CONTENTS OF NUMBER 9

INHALTSVERZEICHNIS HEFT 9

ZEITSCHRIFT FÜR ALLGEMEINE MIKRO-BIOLOGIE

AN INTERNATIONAL JOURNAL ON MORPHOLOGY, PHYSIOLOGY, GENETICS, AND ECOLOGY OF MICROORGANISMS

HEFT 9 · 1982 · BAND 22

AKADEMIE-VERLAG · BERLIN

Die Zeitschrift für Allgemeine Mikrobiologie soll dazu beitragen, Forschung und internationale Zusammenarbeit auf dem Gebiet der Mikrobiologie zu fördern. Es werden Manuskripte aus allen Gebieten der allgemeinen Mikrobiologie veröffentlicht. Arbeiten über Themen aus der medizinischen, landwirtschaftlichen, technischen Mikrobiologie und aus der Taxonomie der Mikroorganismen werden ebenfalls aufgenommen, wenn sie Fragen von allgemeinem Interesse bebehandeln.

Zur Veröffentlichung werden angenommen:

Originalmanuskripte, die in anderen Zeitschriften noch nicht veröffentlicht worden sind und in gleicher Form auch nicht in anderen Zeitschriften erscheinen werden. Der Umfang soll höchstens $1^1/_2$ Druckbogen (24 Druckseiten) betragen. Bei umfangreicheren Manuskripten müssen besondere Vereinbarungen mit der Schriftleitung und dem Verlag getroffen werden.

Kurze Originalmitteilungen über wesentliche, neue Forschungsergebnisse. Umfang im allgemeinen höchstens 3 Druckseiten. Kurze Originalmitteilungen werden beschleunigt veröffentlicht.

Kritische Sammelberichte und Buchbesprechungen nach Vereinbarung mit der Schriftleitung.

Bezugsmöglichkeiten der Zeitschrift für Allgemeine Mikrobiologie:

Bestellungen sind zu richten
- **in der DDR** an eine Buchhandlung oder an den Akademie-Verlag, DDR-1086 Berlin, Leipziger Straße 3–4
- **im sozialistischen Ausland** an eine Buchhandlung für fremdsprachige Literaur oder an den zuständigen Postzeitungsvertrieb
- **in der BRD und Berlin(West)** an eine Buchhandlung oder an die Auslieferungsstelle KUNST UND WISSEN, Erich Bieber, D-7000 Stuttgart 1, Wilhelmstraße 4–6
- **in den übrigen westeuroäpischen Ländern** an eine Buchhandlung oder an die Auslieferungsstelle KUNST UND WISSEN, Erich Bieber GmbH, CH-8008 Zürich, Dufourstraße 51
- **im übrigen Ausland** an den Internationalen Buch- und Zeitschriftenhandel; den Buchexport, Volkseigener Außenhandelsbetrieb der Deutschen Demokratischen Republik, DDR-7010 Leipzig, Postfach 160, oder an den Akademie-Verlag, DDR-1086 Berlin, Leipziger Straße 3–4.

Zeitschrift für Allgemeine Mikrobiologie
Herausgeber: Im Auftrag des Verlages von einem internationalen Wissenschaftlerkollektiv herausgegeben.
Verlag: Akademie-Verlag, DDR-1086 Berlin, Leipziger Straße 3–4; Fernruf: 2236229 oder 2236221
Telex-Nr.: 114420; Bank: Staatsbank der DDR, Berlin, Kto-Nr.: 6836-26-20712.
Chefredaktion: Prof. Dr. UDO TAUBENECK
Prof. Dr. WILHELM SCHWARTZ
Anschrift der Redaktion: Zentralinstitut für Mikrobiologie und experimentelle Therapie der Akademie der Wissenschaften, DDR-6900 Jena, Beutenbergstr. 11; Fernruf: Jena 885614; Telex-Nr.: 058621.
Veröffentlicht unter der Lizenznummer 1306 des Presseamtes beim Vorsitzenden des Ministerrates der Deutschen Demokratischen Republik.
Gesamtherstellung: VEB Druckerei „Thomas Müntzer", DDR-5820 Bad Langensalza.
Erscheinungsweise: Die Zeitschrift für Allgemeine Mikrobiologie erscheint jährlich in einem Band mit 10 Heften. Bezugspreise je Band 250,—M zuzüglich Versandspesen (Preis für die DDR 200,—M). Preis je Heft 25,—M (Preis für die DDR 20,—M).

Erscheinungstermin: November 1982

Bestellnummer dieses Heftes 1070/22/9

AN (EDV) 75218

Zeitschrift für Allgemeine Mikrobiologie	22	9	1982	607-615

(Laboratoire de Technologie Enzymatique, ERA No. 338 du CNRS.
Pr. THOMAS D. B.P. 233 — 60206 Compiegne, France und
Akademie der Wissenschaften der DDR, Zentralinstitut für Mikrobiologie und
experimentelle Therapie, Jena[1], Direktor: Prof. Dr. rer. nat. U. TAUBENECK)

Steroid modifications with immobilized biocatalysts — use of immobilized enzyme-requiring cofactor regeneration and of immobilized mycelium

F. ERGAN, P. ATRAI[1], P. DHULSTER, G. GELLF, M. N. KIM, M. D. LEGOY and D. THOMAS

(Eingegangen am 8. 3. 1982)

Two biological approaches have been investigated for specific modifications of steroids. The first one uses the purified enzyme for the specific dehydrogenation of androsterone to androstanedione. The enzyme used is 3α-hydroxysteroid dehydrogenase which requires a cofactor (NAD). A cofactor regeneration is needed so that the process could work continuously. The conjugation of two points (immobilization of the enzyme and optimization of the ratio methanol-water) allows a continuous work of the enzyme during 25 days. Moreover, we propose a chemical regeneration of the cofactor using methoxy derivative of phenazine methosulphate. Right now it is the limiting step of the process.

The second approach of steroid modification uses a whole mycelium of *Aspergillus phoenicis* for the specific hydroxylation of progesterone to 11α-hydroxy progesterone. The transformation of 90% of the progesterone is obtained with calcium alginate immobilization and the lowest number of products is obtained at pH lower than 2.5 with carrageenan and polyurethane immobilization.

It seems promising to apply immobilized biocatalysts to the bioconversion of hydrophobic compounds in organic solvents system.

Until now little attention has been paid to steroid dehydrogenases. CREMONESI *et al.* (1975) have described the preparation of 20β-hydroxysteroids by using a dehydrogenase. The work of LARSSON and MOSBACH (1976) has involved dehydrogenase in steroid-transforming microorganisms immobilized in polyacrylamide. YAMANE *et al.* (1979) have performed Δ^1-dehydrogenation of steroids in water insoluble organic solvents by free microbial cells and by entrapped cells. YANG and STUDEBAKER (1978) have described the effect of an exogenous electron acceptor (phenazine methosulphate) during the continuous dehydrogenation of steroids by immobilized microbial cells. Steroid dehydrogenation at high substrate concentration using immobilized cells has been performed by CONSTANTINIDES (1980).

On the other hand, studies on microbial steroid conversion have received increasing attention after the discovery by HENCH in 1949 of the cortisone effect on rheumatoids. So 11α-hydroxylation of progesterone (precursor of cortisone) has been well studied: in fact, PETERSON *et al.* (1955) have shown that the product could easily be transformed at a low cost in cortisone. MAMOLI and VERCELLONE (1937) succeeded first in transforming 17 ketosteroids into 17β-hydroxysteroids using a yeast. PETERSON and MURRAY (1952) have performed for the first time the 11α-hydroxylation of progesterone with *Rhizopus nigricans*. Several strains are used by IIZUKA and NAITO (1967) for 11α-hydroxylation, including fungi spores used by VEZINA *et al.* (1963, 1968) and by SINGH *et al.* (1968). OHLSON *et al.* (1979) have immobilized fungi spores but *Aspergillus phoenicis* has not yet been used for this reaction.

The present paper deals first with the continuous use of immobilized 3α hydroxysteroid dehydrogenase (3α HSDH) in water-methanol media, secondly with the use of immobilized *Aspergillus phoenicis*. The common aim is the bioconversion of steroids. Three common limiting factors have been studied in each case:

- In such bioconversions of steroids in organic solvents, immobilization methods and the nature of supports or gels used for immobilization of biocatalysts are important.
- Solvents employed in these transforming systems should be ones which solubilize substrates and products as much as possible and inactive enzymes as little as possible.
- The third limiting factor is the regeneration of cofactor: in the case of 3α HSDH, the cofactor (NAD) is grafted at the active site level of the enzyme and regenerated using an electron carrier.

In the case of steroids hydroxylation the immobilization of whole microorganisms is interesting because it allows the cofactor regeneration.

Materials and methods

3α-hydroxysteroid dehydrogenase (SIGMA, *Pseudomonas testosteroni*) was immobilized with an artificial protein membrane. Membrane production was done by a previously described cross-linking method (LEGOY *et al.* 1980).

Enzyme activity, in the absence of cofactor regeneration was spectrophotometrically measured at 340 nm in a 0.05 M barbital buffer, pH 9.5, solution containing $2.3 \cdot 10^{-3}$ M of NAD. For both native and immobilized enzymes to start the reaction the $3.5 \cdot 10^{-4}$ M androsterone solution in methanol was added in the cuvette. In the presence of cofactor regeneration, 3α-hydroxysteroid-dehydrogenase and NAD cofactor were coimmobilized in the same membrane. No free cofactor was added in the reaction medium. Enzyme activity was then tested with a CLARK pO_2 electrode (E 5040 Radiometer). Cofactor regeneration was performed through a methoxy derivative of phenazine methosulphate (PMS). For continuous stirred tank reactor the same reaction medium oxygenated by air bubbling previously was circulated. In the tank another CLARK pO_2 electrode was used as a reference.

Culture conditions and fermentation: *Aspergillus phoenicis* (strain number 2241 from the Mycotheque Museum Paris) was inoculated (inoculation 1:50 with a 14 hours preculture) into 250 ml flasks each containing 50 ml of the following medium: 40 g malt extract, 3 g yeast extract per liter of water. The medium was shaken in a giratory shaker (180 rpm) at 28 °C–30 °C.

When the evolution of substrate and products concentrations as a function of time was studied, 0.5 ml of a mixture containing 100 μCi of 4^{14}C-progesterone and 1 g of progesterone (Δ^4-pregnene-3,20-dione) in 50 ml of pure ethanol was added to each flask at the end of the growth. The reaction occurred at 30 °C, 160 rpm. In the other cases, cells were separated from the culture medium by simple filtration and the enzyme activity is initiated by the addition of 0.5 ml of the progesterone mixture in 50 ml of buffer. In the range of pH 1.0 to 2.0 hydrochloric acid-potassium chloride buffer 0.2 M was used. In the range of pH 2.5 to 5.0 a citrate-phosphate buffer was used.

Analytical methods: The reaction medium was acidified with 6 ml of 1 M HCl and the products extracted with 20 ml chloroform. The extracts were evaporated to dryness, then redissolved in a small volume of chloroform (200 μl) and chromatographed by thin-layer chromatography on 0.2 mm silica-gel precoated plates in a system chloroform:cyclo-hexane:ethanol (23:23:4 by vol.).

The quantitative estimation of radioactivity after t.l.c. was carried out using a scintillation counter (BECKMANN LS 8000).

Immobilization of cells. Cross-linked albumin foam structure: A solution was made by mixing 9 ml of 0.02 M pH 7 phosphate buffer, 5.6 ml of a 20% albumin solution, 4.5 ml of a 1.5% glutaraldehyde solution and 1.5 g of pellets (wet weight). The mixture was frozen at –20 °C for several hours and then slowly thawed at 4 °C and rinsed.

Cross-linked gelatin foam structure: The same procedure was used after mixing 7.5 ml of phosphate buffer 0.02 M pH 7, 10 ml of a 10% gelatin solution, 2.5 ml of a 1.5% glutaraldehyde solution and 1.5 g of pellets (wet weight).

Alginate and carrageenan gels: 1.5 g of pellets were immobilized, according to previously published procedures inside 3% calcium alginate gel (HACKEL *et al.* 1975, KIERSTAN and BUCKE 1977, OHLSON *et al.* 1979) and inside 2.5% potassium carrageenan (TAKATA *et al.* 1979). Carrageenan were kindly donated by the "Centre de Recherches CECA — F 50500 Carentan Baupté".

Urethan polymers: 1.5 g of pellets were immobilized by mixing with urethan prepolymer PU 3 according to the procedure of FUKUSHIMA *et al.* (1978). Urethan were kindly donated by Professor S. FUKUI (Kyoto University, Japan).

Results and discussion

Use of immobilized enzyme requiring cofactor regeneration for steroid modifications

3α-hydroxysteroid-dehydrogenase catalyses the transformation of androsterone (hydroxysteroid) to 5α-androstan-3,17-dione (ketosteroid) with nicotinamide adenine dinucleotide (NAD) as cofactor.

A systematic study of parameters affecting the activity yield had been done previously (LEGOY *et al.* 1979, 1980).

Due to the low solubility of the substrate, the use of nonaqueous solvent was needed. It was of interest to compare the relative stabilities of free and immobilized enzymes in the presence of pure methanol. A significant increase of the enzyme stability was observed after immobilization. Half life time were 2 and 20 min for free and immobilized enzymes, respectively.

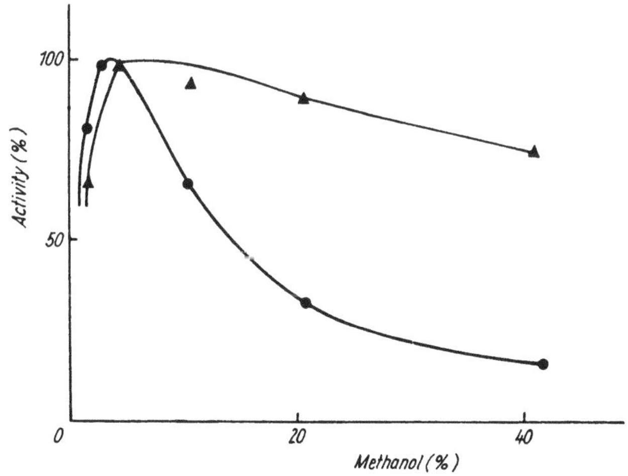

Fig. 1. Activities as a function of methanol concentration in the reaction medium for free (●) and immobilized (▲) enzyme (with fixed substrate concentration)

The activity as a function of methanol concentration in the reaction medium was studied for native and immobilized enzymes (Fig. 1). An increase of the activity was observed until 5%; with higher concentrations, a decrease of the activity was shown. The increase is due to the improvement of the substrate solubility linked to the presence of methanol. The decrease is due to the denaturation of enzyme. The immobilized enzyme is less sensitive than the native one.

The functional stability of immobilized enzyme continuously monitored at 30 °C and the storage stability of free enzyme have been tested (Fig. 2). Immobilized enzyme has worked continuously for 25 days meanwhile the storage stability of the free one was only 16 days.

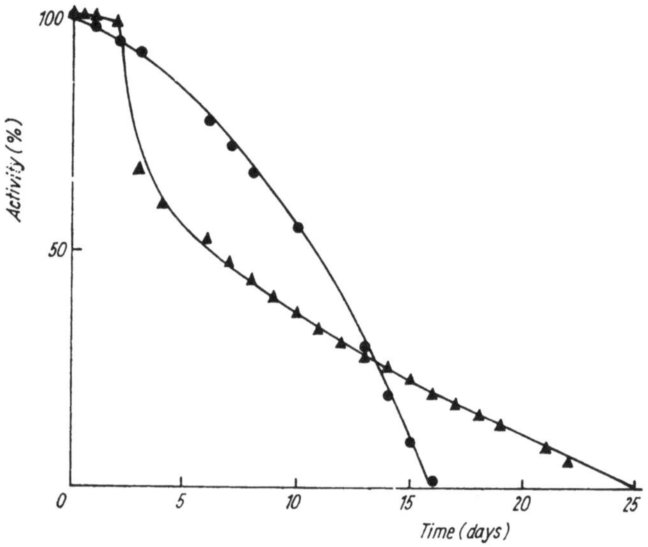

Fig. 2. Resulting enzyme activities as a function of time under continuous use for immobilized enzyme (▲) and in storage conditions for free enzyme (●). Temperature 30 °C

So that the use of 3α HSDH became effective, it was necessary to find a way to eliminate the stoechiometric consumption of the cofactor. Cofactor regeneration had been performed by enzymatic (DAVIES and MOSBACH 1974, GRUNWALD and CHANG 1979), electrochemical (COUGHLIN et al. 1975), and chemical (LEGOY et al. 1980) methods.

In our work, a chemical system of regeneration was used. The enzyme membrane was tested in the absence of free NAD molecules, but in the presence of a methoxy derivative of PMS. The NAD regeneration was performed with oxygen through the methoxy PMS as electron carrier. With a closed batch of 5 cm^3 (Fig. 3) after 30 min the O_2 consumption was nearly total. For a continuously stirred tank reactor (Fig. 4) under the same conditions the steady state was obtained after 3 hours and total inactivation occurred after 90 hours with a flow rate of about 5 $cm^3 h^{-1}$.

Right now, it is necessary to increase the stability of the functional system.

Use of immobilized mycelium for steroid modifications

The bioconversion of progesterone by *Aspergillus phoenicis* producing 11α-hydroxy-progesterone (precursor of cortisone) has been studied. These experiments have been performed with whole cells.

Aspergillus phoenicis transforms progesterone into 11α-hydroxy progesterone and four other products (dihydroxy progesterone, 15β-hydroxy progesterone, 6β-hydroxy progesterone and 16β-hydroxy progesterone).

Progesterone transformation

The concentrations of the substrate and of the 3 most important products (11α-hydroxy progesterone, 15β-hydroxy progesterone, dihydroxyprogesterone) as a function of the reaction time in the culture medium have been studied (Fig. 5). It has been checked that there was no spontaneous degradation of progesterone in the

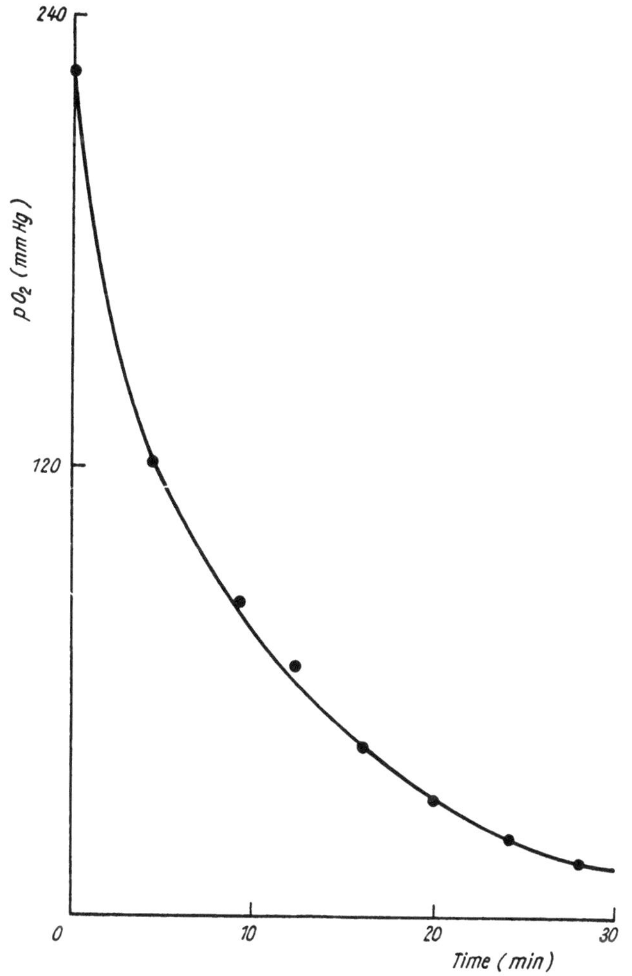

Fig. 3. Oxygen consumption as a function of time during the transient state. Measurements were taken with a pO_2 electrode at 25 °C in a closed batch

medium. After 20 hours of reaction, the transformation reaction is almost total The percentage of progesterone transformed into 11α-hydroxy progesterone presents a maximum after 10 hours of reaction and the percentage of dihydroxy–progesterone increases gradually. 11α-hydroxyprogesterone produced seemed to be transformed in other products as the reaction is going on.

The percentage of progesterone transformed was measured as a function of the reaction time for different culture times: 10, 15, 20 h. (Fig. 6). The optimal culture time is shorter than 10 hours, contrary to what has been observed by MADDOX *et al.* (1981) who found independance of activity and of culture time for *Rhizopus nigricans* immobilized within agar or alginate gels.

To test the effect of reaction pH on the progesterone transformation yield, we used the following buffers: in the range of pH from 1.0 to 2.0 HCl 0.2 M; KCl 0.2 M buffer, in the range of pH from 2.5 to 5.0: citrate-phosphate buffer. The percentage of progesterone transformed as a function of the pH with a culture time of 8 hours and a

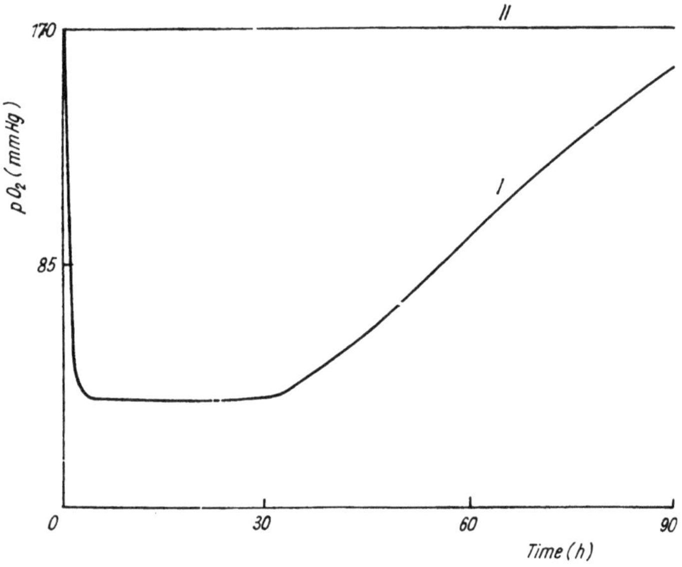

Fig. 4. Oxygen consumption as a function of time for a CSTR. Measurements were taken with a pO_2 electrode in the reactor I, in the tank reserve II

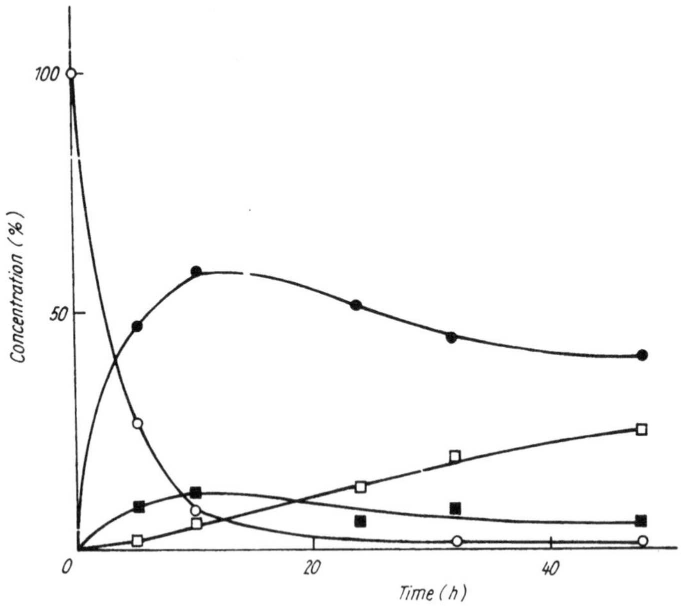

Fig. 5. Concentration of progesterone and of products as a function of time.
○ progesterone untransformed, □ dihydroxyprogesterone, ● 11α-hydroxyprogesterone, ■ 15β-hydroxyprogesterone

reaction medium containing ethanol as the only carbon source showed an optimal pH value around 2.5 and the number of products increases in this range of pH (data not shown). These results were in opposition with SINGH *et al.* (1968) who found that there was no effect of the pH on the transformation activity.

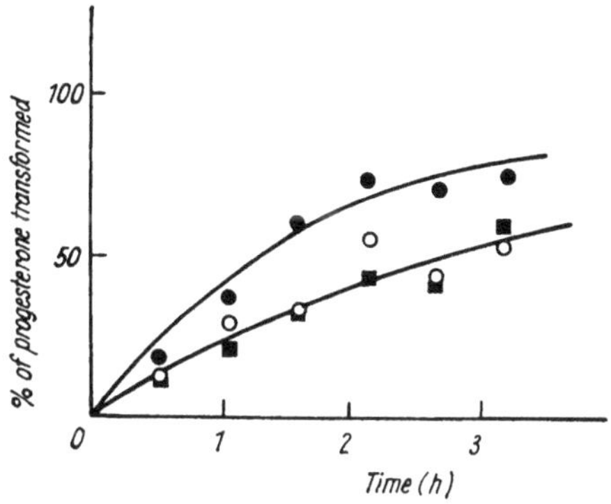

Fig. 6. Percentage of transformed progesterone as a function of reaction time for different culture times: ■ 20 h, ○ 15 h, ● 10 h

Immobilization

Chemicals used for immobilization, the increase of diffusional limitations of the substrate and products modified the enzyme activity. Figures 7, 8 present the percentage of progesterone transformed and of 11α-hydroxy progesterone obtained within the different supports tested. The reaction time was 12 hours. *Aspergillus phoenicis* immobilized within an albumine or a gelatine foam structure had a very low activity. This might be due to the glutaraldehyde used during the immobilization process. Among the different methods of immobilization, calcium alginate was the best for progesterone transformation, but as far as selectivity was concerned, polyurethane and kappa-carrageenan gave better results for 11α-hydroxy progesterone production.

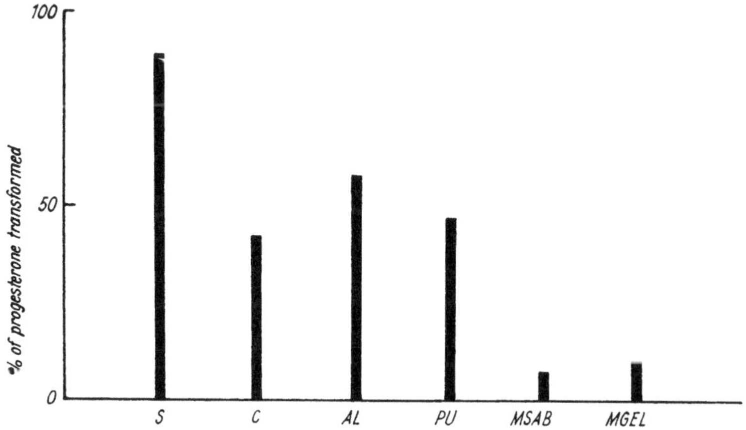

Fig. 7. Percentage of transformed progesterone after 12 hours of reaction when using mycelium immobilized with different methods.
S: Suspension of *Aspergillus phoenicis* in the reaction medium
C: Immobilization of *Aspergillus phoenicis* in Kappa-carrageenan (2.5%)
AL: Immobilization of *Aspergillus phoenicis* in alginate (3%)
PU: Immobilization of *Aspergillus phoenicis* in polyurethan
MSAB: Immobilization of *Aspergillus phoenicis* in an albumin foam
MGEL: Immobilization of *Aspergillus phoenicis* in a gelatin foam

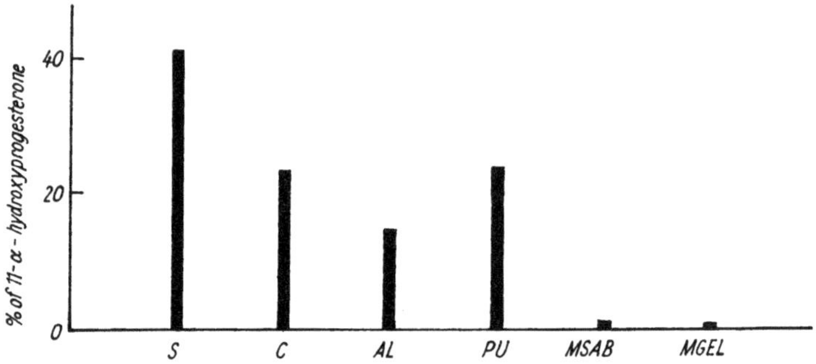

Fig. 8. Percentage of transformed progesterone in 11α-hydroxyprogesterone in the same conditions than Fig. 7

It will be possible to increase the catalysis of progesterone by modifying the culture and reaction conditions. The substrate solubility is a more important problem for steroids transformation because of solvent toxicity. The continuous addition of substrate in the reaction medium with a low solvent concentration might increase the bioconversion of progesterone.

References

CONSTANTINIDES, A., 1980. Steroid transformation at high substrate concentration using immobilized *Corynebacterium simplex* cells. Biotechnol. Bioeng., **22**, 119–136.

COUGHLIN, R. W. and ALEXANDER, B. F., 1975. Simplified flow reactor for electro-chemically driven enzymatic reactions involving cofactors. Biotechnol. Bioeng., **17**, 1379–1382.

CREMONESI, P., CARREA, G., FERRARA, L. and ANTONINI, E., 1975. Enzymatic preparation of 20β-hydroxysteroids in a two phase system. Biotechnol. Bioeng., **17**, 1101–1178.

DAVIES, D. and MOSBACH, K., 1974. The application of immobilized NAD+ in an enzyme electrode and in model enzyme reactors. Biochim. biophysica Acta, **370**, 329–338

FUKUSHIMA, S., NAGAI, T., FUJITA, K., TANAKA, A. and FUKUI, S., 1978. Hydrophilic urethane prepolymers: Convenient materials for enzyme entrapment. Biotechnol. Bioeng., **20**, 1465–1469.

GRUNWALD, J. and CHANG, T. M. S., 1979. Continuous recycling of NAD^+ using an immobilized system of collodion microcapsules containing dextran-NAD^+, alcohol dehydrogenase and malic dehydrogenase. J. Appl. Biochem., **1**, 2, 104–114.

HACKEL, U., KLEIN, J., MEGNET, R. and WAGNER, F., 1975. Immobilisation of microbial cells in polymeric matrices. Eur. J. Appl. Microbiol., **1**, 291–294.

IIZUKA, H. and NAITO, A., 1967. Microbial Transformation of Steroids and Alkaloids. University of Tokyo Press.

KIERSTAN, M. and BUCKE, C., 1977. The immobilization of microbial cells, subcellular organelles, and enzymes in calcium alginate gels. Biotechnol. Bioeng., **19**, 387–397.

LARSSON, P. O. and MOSBACH, K., 1976. Immobilization of steroid-transforming microorganisms in polyacrylamide. In: Methods in Enzymology, Vol. 44. Academic Press, New York.

LEGOY, M. D., LARRETA GARDE, V., ERGAN, F. and THOMAS, D., 1979. Study of immobilized α steroid dehydrogenase in water-methanol media. J. Solid-Phase Biochem., **4**, 2, 143–148.

LEGOY, M. D., LARRETA GARDE, V., LE MOULLEC, J. M., ERGAN, F. and THOMAS, D., 1980. Cofactor regeneration in immobilized enzyme systems: Chemical grafting of functional NAD in the active site of dehydrogenases. Biochimie, **62**, 341–345.

MADDOX, I. S., DUNNILL, P. and LILLY, M. D., 1981. Use of immobilized cells of *Rhizopus nigricans* for the 11α hydroxylation of progesterone. Biotechnol. Bioeng., **23**, 345–354.

MAMOLI, L. and VERCELLONE, A., 1937. Z. Physiol. Chem., **245**, 93–101.

OHLSON, S., LARSSON, P. O. and MOSBACH, K., 1979. Steroid transformation by living cells immobilized in calcium alginate. Eur. J. Appl. Microbiol. Biotechnol., **7**, 103–110.

PETERSON, D. H., MEISTER, P. D., WEINTRAUB, A., REINEKE, L. M., EPPSTEIN, S. H., MURRAY, H. C. and LEIGH OSBORN, H. M., 1955. Microbiological transformation of steroids. XIII. Oxy-

genation of 16α, 17α-Oxidoprogesterone to 11α-Hydroxy-16α, 17α-Oxidoprogesterone by *Rhizopus nigricans* EHRB. (ATCC 62276). J. Amer. chem. Soc., **77**, 4428—4435.
PETERSON, D. H. and MURR Y, H. C., 1952. Microbial oxygenation of steroids and carbon 11. J. Amer. chem. Soc., **74**, 1872.
SINGH, K., SEHGAL, S. N. and VEZINA, C . 1968. Large-scale transformation of steroids by fungal spores. Appl. Microbiol., **16**, 393—397.
TAKATA, I., YAMAMOTO, K., TOSA, T. and CHIBATA, I., 1979. Screening of microorganisms having high fumarase activity and their immobilization with carrageenan. Eur. J. Appl. Microbiol. Biotechnol., **7**, 161—172.
VEZINA, C., SEHGAL, S. N. and SINGH, K., 1963. Transformation of steroids by spores of microorganisms. I. Hydroxylation of progesterone by conidia of *Aspergillus ochraceus*. Appl. Microbiol., **11**, 50—57.
VEZINA, C., SEHGAL, S. N. and SINGH, K., 1968. Transformation of organic compounds by fungal spores. Adv. Appl. Microbiol., **10**, 221—268.
YAMANE, T., NAKATANI, H. and SADA, E., 1979. Steroid bioconversion in water-insoluble organic solvents: Δ^1-dehydrogenation by free microbiol cells and by cells entrapped in hydrophilic or lipophilic gels. Biotechnol. Bioeng., **29**, 2133—2145.
YANG, H. S. and STUDEBAKER, J. F., 1978. Continuous dehydrogenation of a steroid with immobilized microbial cells: effect of an exogenous electron acceptor. Biotechnol. Bioeng., **20**, 17—25.

Mailing address: Dr. F. ERGAN
Laboratoire de Technologie Enzymatique, ERA No. 338 du CNRS
60206 Compiegne/Cedex, France

Zeitschrift für Allgemeine Mikrobiologie	22	9	1982	617–627

(Akademie der Wissenschaften der DDR,
Forschungszentrum für Molekularbiologie und Medizin,
Zentralinstitut für Mikrobiologie und experimentelle Therapie, Jena,
Direktor: Prof. Dr. U. TAUBENECK)

Growth characteristics and ultrastructure of protoplast type L-forms from Streptomycetes

J. GUMPERT

(Eingegangen am 18. 12. 1981)

L-form colonies from *S. hygroscopicus*, *S. griseus* and *S. levoris* were isolated after incubation of lysozyme protoplasts on an osmotically stabilized complex agar medium. Unstable and stable L-forms grow on solid and in liquid media. L-form colonies are 5–10 times smaller than normal colonies and show a typical morphology for each species. In ultrathin sections L-form cells are characterized by nucleoid areas with typical core-like structures, by a ribosome-rich cytoplasm with different inclusion bodies, and by a cytoplasmic membrane. Because there are no cell wall structures L-forms of the three *Streptomyces* species belong to the protoplast type.

Analysis of cell size and cell shape shows a variation in diameter and a cell propagation by regular and irregular division- and budding-like processes. Many L-form cells contain more than two chromosomes.

The results are discussed with regard to the cellular organisation of streptomycetes and the nature of the stable L-form.

Although much work has been done about protoplasts of streptomycetes (DOUGLAS *et al.* 1958, HOPWOOD *et al.* 1977, KOCHKINA and RAUTENSTEIN 1968, OKANISHI *et al.* 1966, 1974, SAGARA *et al.* 1971) no data exist about *Streptomyces* L-forms, *i.e.* about growing and dividing protoplasts or spheroplasts of these organisms.

Cells of streptomycetes growing and propagating in the protoplast state as L-forms might be of great interest for several reasons: they offer the possibility of obtaining *Streptomyces* cells more uniform in age and physiological state, from which cellular components can be isolated more easily; they seem to be useful objects to study physiological and biochemical questions, especially the biosynthesis of primary and secondary metabolites; and they are good candidates for genetic experiments, particularly for fusion, tranformation and transfection processes.

In two papers (BAUDLER and GUMPERT 1979, GUMPERT 1978) we have described the isolation of *Streptomyces hygroscopicus* L-form. In this report colony morphology, growth characteristics, and ultrastructure of L-forms from three *Streptomyces* species are summarized.

Materials and methods

Organism: *Streptomyces hygroscopicus* NG 33-354 (IMET Jena 40668), *S. griseus* (IMET JA 5142), and *S. levoris* (EK 1331) are strains from the collection of the Zentralinstitut für Mikrobiologie und experimentelle Therapie Jena.

Spores conserved in earth (LUVOS Heilerde) were used for maintaining the strains.

Induction of L-form growth: *Streptomyces* strains were grown for 16–20 h at 28 °C in a complex medium composed of mineral salts, glucose (1%), peptone (0.5%), yeast extract (0.4%), and glycine (1%). Mycelia are transformed into protoplasts by lysozyme treatment (2000 µg/ml) in protoplasting medium according to OKANISHI *et al.* (1974).

Protoplasts were purified by filtration through a glas filter (VEB SCHOTT, Jena, GDR, pore diameter about 20 µm) and inoculated onto an L-induction medium (LIM, consisting of 0.4 M saccharose, 0.06 M glucose, 1.4 mM K_2SO_4, 0.37 mM KH_2PO_4, 35 mM $CaCl_2 \cdot 6\,H_2O$, 3% yeast extract (DIFCO), 0.025 M TES buffer, 5% horse serum, 2% glycine, 1000 U/ml penicillin, and 1% agar). L-form colonies appear after 5–8 d incubation at 28 °C under aerobic conditions.

L-form cultivation on solid media: For cultivation on agar medium small agar blocks with abundant growth of L-form colonies were cut out from plates, the colony side layed on the agar surface of a new plate of L-form culture medium (LCM) and moved across the agar surface. Before inoculation plates were dried for 30 min at 37 °C. LCM has the following composition: 0.4 M saccharose, 0.06 M glucose, 1.4 mM K_2SO_4, 0.37 mM KH_2PO_4, 20 mM $MgCl_2 \cdot 6\,H_2O$, 35 mM $CaCl_2 \times 6\,H_2O$, 7 mM $(NH_4)_2SO_4$, 13 mM NH_4NO_3, 1% yeast extract, 0.025 M TRIS-buffer pH 7.2, 2 ml/1000 ml trace element solution. Penicillin G (1000 U/ml) and glycine (1–2%) were added to prevent cell wall regeneration. In first experiments 0.4% vitamin-free casamino acids (DIFCO) and TES buffer were used. All supplements marked with an asterisk were sterilized separately and added before preparing the plates. Transfer onto fresh medium was carried out in intervals of 8–14 d.

For isolating stable L-forms a few or single colonies were transferred in parallel on LCM with and without penicillin and glycine supplements. All cultures were incubated at 28 °C.

L-form cultivation in liquid media: LCM was used for adaptation of L-forms to growth in liquid media. Agar blocks with abundant L-form growth were put into liquid LCM and incubated at 28 °C. L-form growth could also be obtained in liquid medium by successively lowering the agar concentration from 1 to 0.4% in the course of several passages.

Light microspopy: A stereoscopic microscope SM XX (VEB C. ZEISS, Jena, GDR) was used for investigating colonies. Microcolonies and cell suspensions were examined using an Amplival microscope equipped with phase contrast and with a microphotographic unit mf-matic (VEB C. ZEISS, Jena GDR).

Electron microscopy: Small agar blocks with L-form colonies grown for 6–10 d on LCM were cut out from plates and fixed in glutaraldehyde (5% in solution B, 16 h, 4 °C), washed in solution B, fixed in OsO_4 (1% in solution B, 4.5 h 20 °C), washed again in solution B, dehydrated in acetone, and embedded in Mikropal (FLUKA). Solution B is a Na-acetate veronal buffer supplemented with HCl and $CaCl_2$ according to RYTER and KELLENBERGER (1958).

Ultrathin sections were made by using an Ultrotome III (LKB). They were poststained with lead citrate according to REYNOLDS (1963) and investigated in a SIEMENS Elmiskop I operating at 80 kV.

Results

L-form growth on solid media

When a purified protoplast suspension of about 10^8–10^9 cells/ml was distributed by spatule on predried LIM agar plates and incubated at 28 °C, very small punctiform colonies appeared after 7–10 d consisting exclusively of spherical cells (Fig. 1a). In the course of subsequent transfers on fresh LCM agar colonies become larger and show a characteristic morphology (Figs. 1, 2, 3). Simultaneously growth becomes faster and after some passages colonies can be recognized after 4–5 d.

In the course of further cultivation L-form growth could also be obtained on agar media without serum and supplemented only with vitamin-free casamino acids or with a mixture of 15 amino acids according to MEDILL and O'KANE (1954) as nitrogen sources.

Unstable L-forms produce several or many reverting colonies on media without glycine and penicillin supplements. These colonies consist of a mixture of L-form cells and regenerated hyphal cells or of hyphae only (Fig. 4). Stable L-forms, however, grow without regeneration on LCM agar not supplemented with agents inhibiting cell wall biosynthesis and preventing regeneration. These cells have lost the capacity to resynthesize their walls.

Colony morphology and growth characteristics differ among *Streptomyces* species. Their properties are summarized in Table 1 and in Figs. 1, 2 and 3. Stable L-form

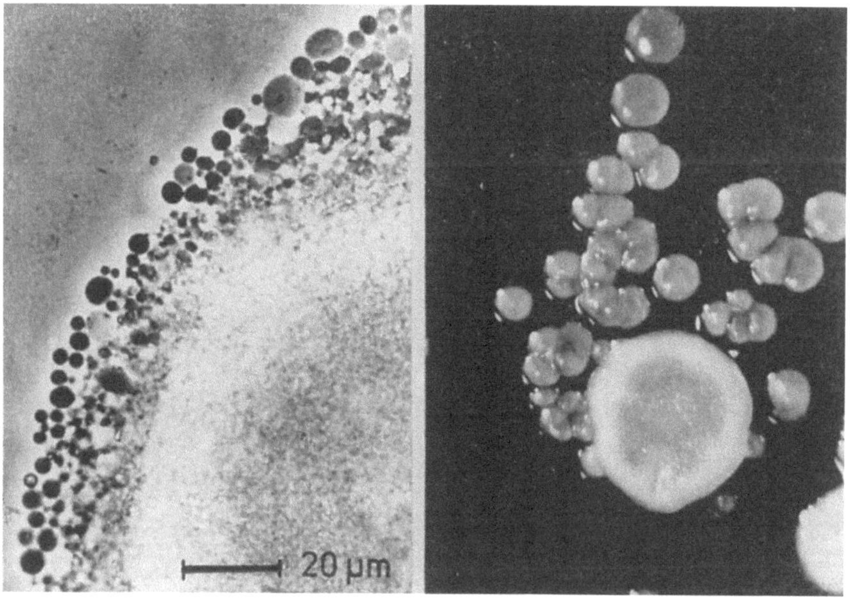

Fig. 1. L-form colonies of *Streptomyces griseus*. 10 d old.
a) margin of a colony showing spherical L-form cells.
b) L-form colonies without a fried egg appearance. The large colony is a revertant consisting mainly of walled hyphal cells (10:1)

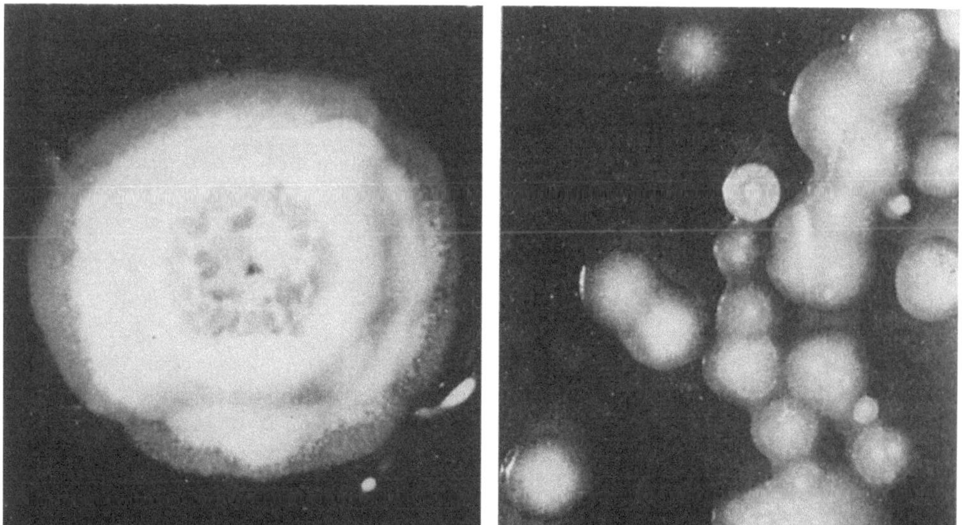

Fig. 2. Normal colony, 10 d old, (a, 1:6) and L-form colonies, 14 d old, (b, 1:8) of *Streptomyces hygroscopicus*

cultures from *S. hygroscopicus* and *S. levoris* appeared after 4—10 successive transfers on penicillin-glycine medium.

Until now it was not possible to isolate stable L-forms from *S. griseus* in the same way. When colonies of this strain grown more than 50 passages in the unstable L-form

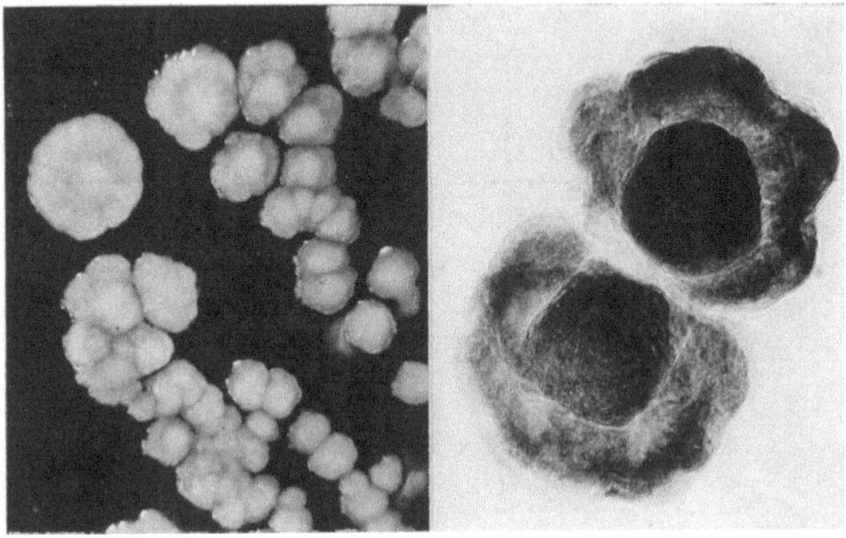

Fig. 3. L-form colonies, 12 d old, of *Streptomyces levoris* showing characteristic fried egg appearance. a) 1:10, b) phase contrast micrograph 1:60

Table 1

Characteristics of L-form growth in *Streptomyces hygroscopicus*, *S. levoris* and *S. griseus*

	S. hygroscopicus	*S. levoris*	*S. griseus*
colony morphology	1—4 mm ∅ circular, raised or convex, smooth, fried egg appearance, with a small dense center and a more translucent peripheric zone containing yellowish or white inclusions and radiated structures	0.5—3 mm ∅ circular with undulated margins, raised or convex, opaque, smooth, grayish to brown, fried egg appearance, with a large dense center, formation of crystalline inclusions after 10 d incubation	0.5—1.5 mm ∅ punctiform or circular, translucent, smooth, grayish white, no typical fried egg appearance
regeneration on LCM supplemented with:			
400 U/ml penicillin and 2% glycine	absent	absent	absent
2% glycine	rare	rare	frequent
0.2% glycine	frequent	frequent	complete
isolation of stable L-form	+ more than 100 transfers without regeneration	+ more than 50 transfers without regeneration	—

state are transferred to penicillin- and glycine-free media many reverting colonies appear on the plates.

L-form growth in liquid media

When agar blocks with abundant L-form colonies were put in small bottles with liquid LCM, growth of L-form cells in sediment and in surface layer could be obtained. After several passages yields of about 10^9 cells/ml were reached by incubating 20—50 ml

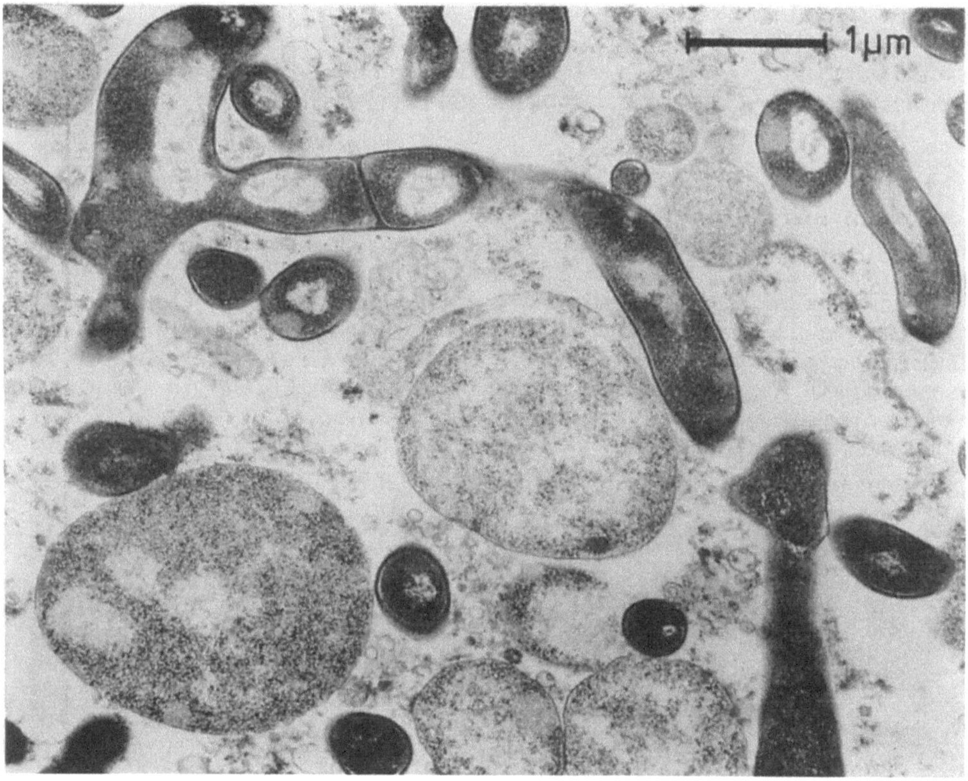

Fig. 4. Hyphal cells and L-form cells in a reverting unstable L-form colony of *Streptomyces hygroscopicus*

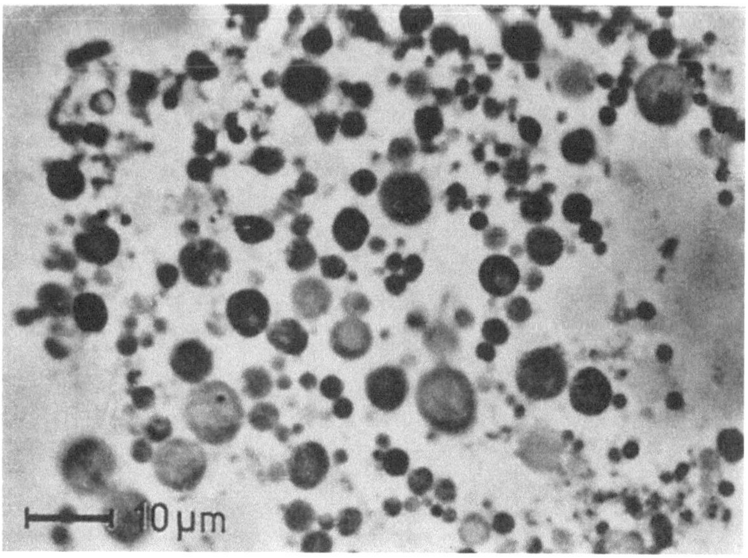

Fig. 5. L-form cells of *Streptomyces hygroscopicus* grown in liquid medium

medium for 8 d at 28 °C without shaking. Aggregates of spherical cells varying in diameter from 0.5—3 µm are typical (Fig. 5). Also single, very large cells of 5—10 µm can be seen. Many L-form cells contain vacuoles or areas of lower contrast indicating structural changes and compartmentation within the cell contents.

Ultrastructure of L-form cells

Spherical and pleomorphic cell forms of different size are to be seen in ultrathin sections from L-form colonies (Fig. 6, 7 and 9). Cell diameters vary between 0.5 and 6 µm. Pleomorphic cells sometimes reach dimensions of about 10 µm.

Besides typical cells smaller and larger accumulations of vesicular structures, 0.05—0.4 µm in diameter, occur (Fig. 6, 8). Obviously vesicles are produced by different ways. Serial sections show that small vesicles, mostly filled with electron dense material, are produced at the surface of intact cells. On the other hand similar vesicles are produced within destructively changed cells.

The analysis of cell shape allows conclusions about the mode of L-form propagation. Besides a more or less regular cell division budding-like processes occur. Daughter

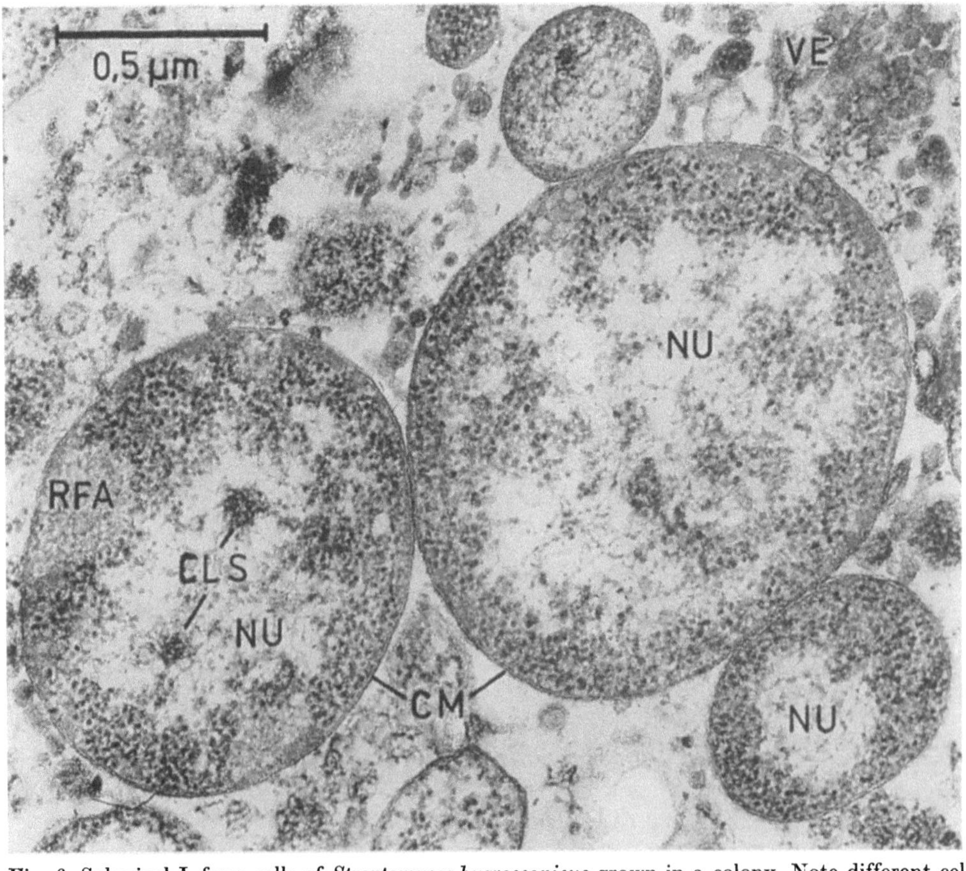

Fig. 6. Spherical L-form cells of *Streptomyces hygroscopicus* grown in a colony. Note different cell diameters, cytoplasmic membranes (CM) as the only cell envelope structure, ribosome-free areas (RFA), electron dense core-like structures (CLS) in nucleoid area (NU), and accumulation of small vesicles (VE)

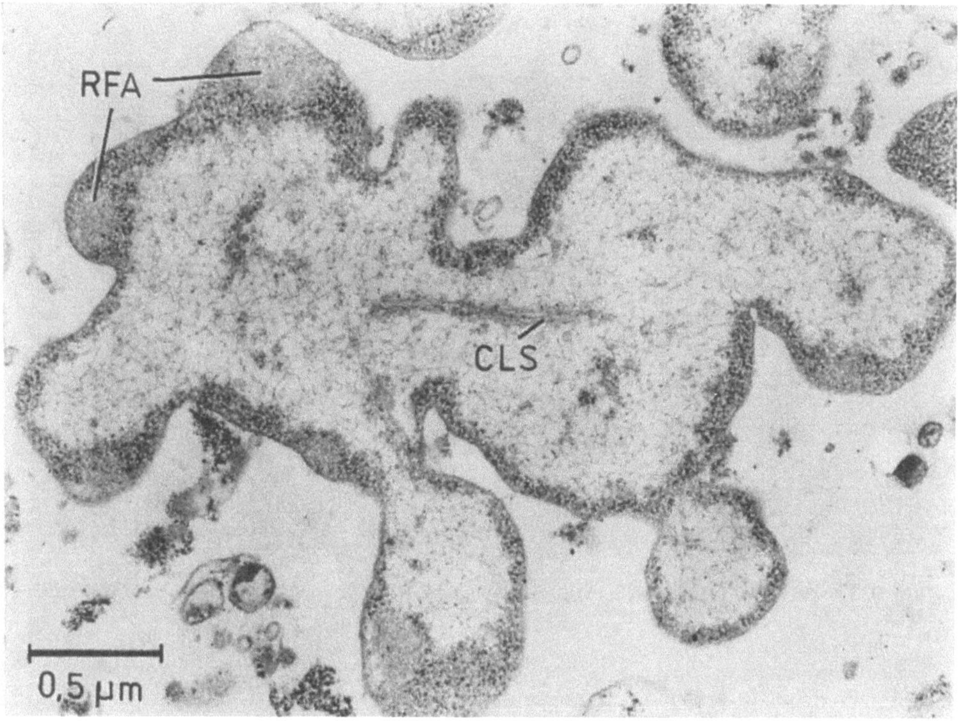

Fig. 7. Pleomorphic L-form cell of *Streptomyces levoris*. Note the electron dense core-like structure (CLS) in nucleoid area and budd-like proliferations. RFA: ribosome-free areas

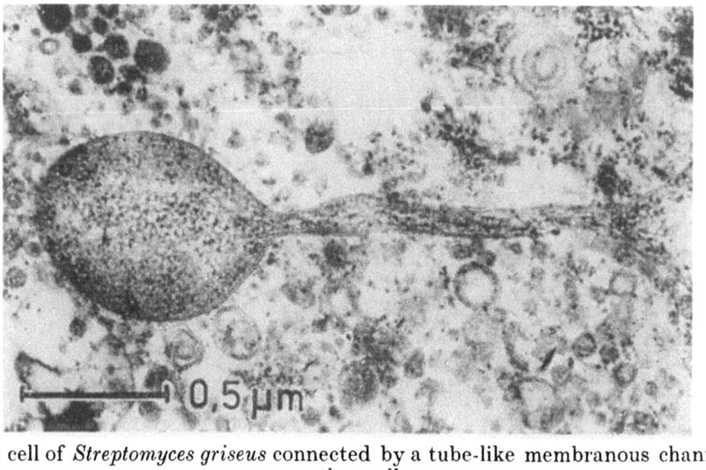

Fig. 8. L-form cell of *Streptomyces griseus* connected by a tube-like membranous channel with the mother cell

cells are formed which grow in close connection with mother cells. They arise from the surface of a mother cell (Fig. 7) or in a distance of more than 1 µm (Fig. 8, 9). In most cases they are smaller than mother cells after separation.

L-forms of the three *Streptomyces* species are of protoplast type. Cells are surrounded by a typical trilamellar cytoplasmic membrane. No remnants of cell wall structures can be recognized (Fig. 6, 7, 8, 9).

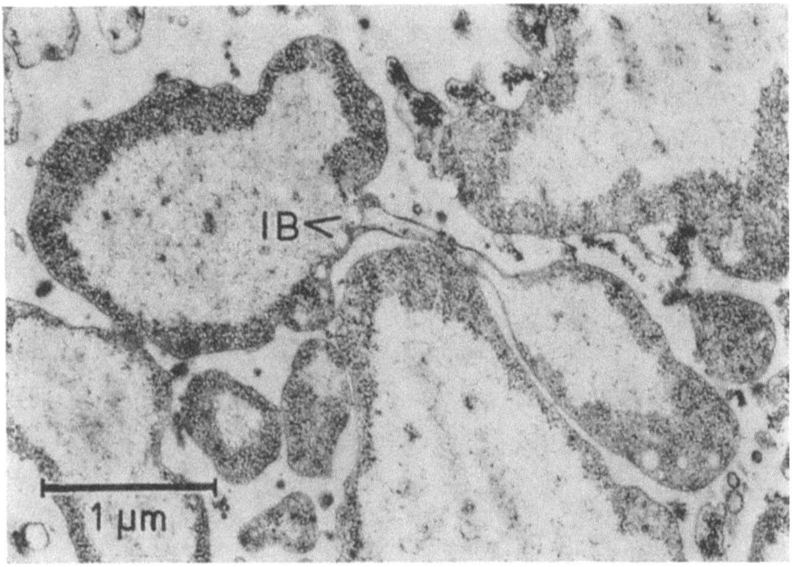

Fig. 9. Propagating L-form cells of *Streptomyces griseus* connected by a membranous tube.
IB: inclusion bodies

A ribosome-rich cytoplasm and more or less extended nucleoid regions can be distinguished inside the cells. Because of the glutaraldehyde fixation nucleoid areas are often different and do not correspond to the nucleoid shape *in vivo*.

Characteristic electron dense material with fibrils occurs in the center of many nucleoid areas (Fig. 6, 7). Obviously, these structures are not simply inclusions of cytoplasmic material but seem to be typical nucleoid structures. They might represent those core structures proposed by WORCEL and BURGI (1972) for the folded bacterial chromosome.

Characteristic medium contrasted ribosome-free areas occur regularly in the cytoplasm and preferentially at the cytoplasmic membrane (Fig. 6, 7). Their size varies from 0.05 – 0.8 μm. In lysed cells the material can remain attached to the cytoplasmic membrane. These ribosome-free areas may be considered as inclusion bodies of homogeneous but unknown material.

In *S. griseus* inclusion bodies of another type occurred regularly. They are characterized by a regular circular shape, by a completely electron opaque interior, and by a sharp demarcation against the cytoplasm, which is not a membrane (Fig. 9). They are apparently glycolipid granules which also occur in normal hyphal cells of streptomycetes.

There are several ultrastructural peculiarities such as unusual alterations in cell surface, formation of multiple extracellular membranes, intracellular crystalline inclusion bodies, and giant core structures. They will be described and discussed in another paper.

Discussion

Conversion of bacterial cells into protoplast or spheroplast type L-forms is well-known for many Gram-positive and Gram-negative species (HIJMANS *et al.* 1969, MATTMAN 1974).

In actinomycetes L-form growth has only been reported for *Nocardia asteroides* (BURGEOIS and BEAMAN 1976) and *Agromyces ramosus* (HORWITZ and CASIDA 1975).

In comparison to L-form induction in other bacteria *Streptomyces* L-form growth could only be obtained if isolated protoplasts were used as inoculum. All attempts to induce L-form growth by treating normal cells with lysozyme, penicillins, or other cell wall inhibiting agents, which is the usual method for L-form induction, were without success.

Although L-form colonies of the three *Streptomyces* species used in these experiments consist of cell wall-less protoplast cells of the same morphological appearance, each species shows characteristic differences in colony morphology (Fig. 1, 2, 3). Obviously specific properties in growth and division processes, in cell arrangement, and in cell surface behaviour exist in each species which are responsible for the typical colony morphology.

The general ultrastructure of L-form cells is similar to those of protoplasts derived from the parent strains or from other *Streptomyces* species. Variation in size, absence of cell wall structures and mesosomes, more or less extended nucleoid regions with core-like structures, ribosome-rich cytoplasm with inclusion bodies, and formation of vesicles at the surface have also been observed in protoplasts of various other streptomycetes (GUMPERT 1978, 1980, OKANISHI *et al.* 1974, SAGARA *et al.* 1971). The different amount of vesicular structures in various regions of a L-form colony shows that they are not mesosomal structures, and that their formation depends on the physiological state of the cells and on environmental conditions. Some of the vesicles are formed inside the cell as the result of degradative processes. They are released into the environment after rupture of the cytoplasmic membrane. Vesicles which are formed at the surface of living L-form cells are more interesting because cytoplasmic material and high molecular-weight substances can be removed from the cell by that way.

The growth of *Streptomyces* as spherical and pleomorphic wall-less cells allows some conclusions about their cellular organisation. The smallest vegetative cell is a hyphal unit. It is a hyphal part several µm in length and demarcated from others by irregular cross walls which do not split and separate the cells. These hyphal units contain numerous chromosomes in a common cytoplasm surrounded by a common cytoplasmic membrane. Light microscopic measurements of the hyphal length divided by the number of nucleoids contained in it revealed values of 1.4–1.9 µm for a unit cell (GUMPERT, unpubl., SCHUHMANN and BERGTER 1976). This corresponds to the bacterial unit cell postulated by DONACHIE and BEGG (1970) for the volume of one resting chromosome and the corresponding amount of cytoplasm. A large portion of growing L-form cells as well as of newly formed protoplasts during lysozyme treatment (GUMPERT, unpubl.) corresponds in volume to 1–2 unit cells. This fact demonstrates that streptomycetes are able to grow not only as polyenergidic hyphal cells but also as monoenergidic cells like most other bacteria. It supports further the conception that the hyphal units consist of several of such unit cells which functionally, and in a certain sense also structurally represent autonomous entities. Chromosomes may replicate and divide within a polyenergidic hyphal unit independently from those of the adjacent unit cells.

Not all L-form cells grow as monoenergidic cells. The variation in cell volume and cell shape and the ultrastructure of the propagation stages show that normal cell division is frequently disturbed. Many cells contain more than 2 nucleoids, and obviously replication of chromosomes is not followed by subsequent segregation and separation processes. Many large cells form smaller daughter cells by budding-like processes (Fig. 7, 8, 9). The formation of a daughter cell at the end of a membranous channel seems to occur only in colonies growing on agar media. Obviously internal and external factors are responsible for those irregular division processes, *e.g.* very slow replication and incomplete segregation of chromosomes, changed membrane properties, plasticity of cells, consistency and fibrillar structure of the surrounding agar gel, and cell density in colonies.

Whereas isolation of stable L-form colonies was easy with *S. hygroscopicus* and *S. levoris*, the efforts with *S. griseus* have been negative till now. The reasons must lie in the nature of the stable *Streptomyces* L-form, however, more biochemical and genetic investigations are necessary to explain it precisely.

The loss of cell wall biosynthesis under normal growth conditions is such a distinct and stable marker that stable L-forms may be considered as mutants. However, the fact that they arise without agents interacting with DNA and that a large number of colonies becomes stable after a few transfers on penicillin or glycine-supplemented media speaks against a primary mutative event.

It seems more likely that defects in gene expression or in regulation of wall biosynthesis cause this stability. Bacterial wall polymers are made from soluble nucleotide intermediates by membrane enzymes (ROGERS 1970). As consequences of membrane alterations and of the absence of the periplasmic space, an important milieu for various processes, the coordinated transport of murein- and other nucleotides across the membrane and the localization and action of enzymes involved in construction of the murein and teichoic acid network may be disturbed. Such a non-mutative stability does not exclude the possibility that L-form types originate in the course of the thousands of generations during subsequent cultivation which have lost the primary genetic information for enzymes or building blocks involved in cell wall biosynthesis and which can be considered to be true mutants.

Acknowledgement

I would like to thank Mrs. MONIKA VÖLKEL for her excellent technical help.

References

BAUDLER, E. and GUMPERT, J., 1979. Isolation of a protoplast type L-form from *Streptomyces hygroscopicus*. Z. Allg. Mikrobiol., **19**, 363–365.

BOURGEOIS, L. and BEAMAN, B. L., 1976. *In vitro* spheroplast and L-form induction within the pathogenic nocardiae. J. Bacteriol., **127**, 584–594.

DONACHIE, W. D. and BEGG, K. J., 1970. Growth of the bacterial cell. Nature (London), **227**, 1220–1224.

DOUGLAS, R. J., ROBINSON, J. B. and CORKE, C. T., 1958. On the formation of protoplast-like structures from streptomycetes. Canad. J. Microbiol., **4**, 551–554.

GUMPERT, J., 1978. Ultrastructure and modifiability of the cell envelope in *Streptomyces hygroscopicus*: In: *Nocardia* and *Streptomyces*, Proceedings of the International Symposium on *Nocardia* and *Streptomyces*. G. Fischer Verlag Stuttgart–New York, 221–233.

GUMPERT, J., 1980. Electron microscopic analysisof protoplast fusion in *Streptomyces hygroscopicus* and consideration on structural alterations in fusing membranes. Arch. Microbiol., **126**, 263–269.

HIJMANS, W., VAN BOVEN, C. P. A. and CLASENER, H. A. L., 1969. Fundamental biology of the L-phase of bacteria: In: HAYFLICK, L., The Mycoplasmatales and the L-phase of Bacteria. North Holland Publ. Comp. Amsterdam, 67–143.

HOPWOOD, D. A., WRIGHT, H. M., BIBB, M. J. and COHEN, S. N., 1977. Genetic recombinations through protoplast fusion in *Streptomyces*. Nature (London), **268**, 171–174.

HORWITZ, A. H. and CASIDA, jr. L. E., 1975. L-phase variants of *Agromyces ramosus*. Antonie van Leeuwenhoek, **41**, 153–171.

KOCHKINA, J. M. and RAUTENSTEIN, Y. I., 1968. Formation of protoplasts in *Actinomyces streptomycini* KRASS (in Russian). Mikrobiologija, **37**, 1046–1051.

MATTMAN, L., 1974. Cell Wall Deficient Forms. CRC Press Cleveland.
MEDILL, M. A. and O'KANE, D. J., 1954. A synthetic medium for the L-type colonies of *Proteus*. J. Bacteriol., **68**, 530–533.
OKANISHI, M., UTAHARA, R. and OKAMI, Y., 1966. Infection of the protoplasts of *Streptomyces kanamyceticus* with deoxyribonucleic acid preparation from actinophage PK-66. J. Bacteriol., **92**, 1850–1852.
OKANISHI, M., SUZUKI, K. and UMEZAWA, H., 1974. Formation and reversion of streptomycete protoplasts: cultural condition and morphological study. J. Gen. Microbiol., **80**, 389–400.
REYNOLDS, E. S., 1963. The use of lead citrate at high pH as an electron opaque stain in electron microscopy. J. Cell. Biol., **32**, 27–31.
ROGERS, H. J., 1970. Bacterial growth and the cell envelope. Bact. Rev., **34**, 194–214.
RYTER, A. et KELLENBERGER, E., 1958. Etude au microscope électronique de plasmas contenant de l'acide deoxyribonucleique. Z. Naturforsch., **13b**, 597–605.
SAGARA, Y., FUKUIK, K., OTA, F., YOSHIDA, N., KASHIYAMA, T. and FUJIMOTO, M., 1971. Rapid formation of protoplasts of *Streptomyces griseoflavus* and their fine structure. Jap. J. Microbiol., **15**, 73–84.
SCHUHMANN, E. und BERGTER, F., 1976. Mikroskopische Untersuchungen zur Wachstumskinetik von *Streptomyces hygroscopicus*. Z. Allg. Mikrobiol., **16**, 201–215.
WORCEL, A. and BURGI, E., 1972. On the structure of the folded chromosome of *Escherichia coli*. J. molecular Biol., **43**, 127–147.

Mailing address: Dr. J. GUMPERT
Zentralinstitut für Mikrobiologie
und experimentelle Therapie der AdW
DDR-6900 Jena, Beutenbergstr. 11

Zeitschrift für Allgemeine Mikrobiologie	22	9	1982	629–637

(Akademie der Wissenschaften der DDR,
Forschungszentrum für Molekularbiologie und Medizin,
Zentralinstitut für Mikrobiologie und experimentelle Therapie, Jena,
Direktor: Prof. Dr. U. TAUBENECK)

Alteration of interaction with virulent bacteriophage Ta1 during differentiation of *Thermoactinomyces vulgaris*

SIGRID KRETSCHMER

(Eingegangen am 10. 3. 1982)

During its life cycle *Thermoactinomyces vulgaris* 1227 and 1261 substrate mycelium showed three modes of interaction with the virulent phage Ta1, each expressed at a distinct morphological stage. Primary mycelium arising from spores 50 min after start of germination was the only stage which propagated phages upon infection. However, infection of growing secondary mycelium or of late sporulation stages resulted in a loss of phage. The lack of phage production upon infection of growing secondary mycelium was not related to sporogenesis, since it appeared already 3.5 hours before the beginning of sporulation. If phages were added to the secondary mycelium at the beginning of spore formation, the phage genome became integrated in the developing spores in a heat-stable state. Allowing outgrowth of these prophage-carrier spores, phages were produced similarly as in germ tubes arising from normal spores infected at the time of inoculation.

The growing secondary substrate mycelium was characterized by competence for the uptake of exogenous DNA. Since at the stage of competence phages were neither produced nor was phage DNA trapped, the earlier reported lack of transfection in undisturbed differentiating *T. vulgaris* is now understandable.

In contrast to *Bacillus subtilis* where competent cells are susceptible to both transformation and transfection (GREEN 1964, REILLY and SPIZIZEN 1965, KRETSCHMER and SARFERT 1971), in *Thermoactinomyces vulgaris* transfection did not occur under conditions favourable for transformation. Plaques were found only if a modified technique was used, by which obviously the life cycle of *T. vulgaris* 1227 was interrupted and resumption of vegetative growth was induced (KRETSCHMER and SARFERT 1980). The plaque forming units (p.f.u.) found using the improved transfection method were either free phages or prophage-carrier spores.

B. subtilis and *T. vulgaris* are both differentiating procaryotes, and their mode of endospore formation is similar (DOROKHOVA *et al.* 1970, CROSS *et al.* 1971). Their life cycles, however, differ at least by the existence of a secondary growth phase in *T. vulgaris*, separating primary growth from sporulation (KRETSCHMER 1978a). Asking for possible reasons why competent cells of *B. subtilis* and *T. vulgaris* respond differently to phage DNA infection, this developmental heterology may be taken into consideration.

The present study was undertaken to correlate morphologically different stages of the life cycle of *T. vulgaris* substrate mycelium to the following transiently expressed abilities: to undergo transformation, to produce phages and to trap phage DNA into developing spores.

Materials and methods

Bacteria and phage strains: The competent strain *Thermoactinomyces vulgaris* 1227 and its mutant derivative 1261 *nic-1 str-1 thi-3* were kindly supplied by D. A. HOPWOOD. From strain 1261 a mutant was isolated which is defective in the formation of aerial mycelium (amy^-) but not impaired in its ability to sporulate.

The virulent phage most frequently employed was Ta1 which along with its DNA ($28.8 \cdot 10^6$ daltons/phage) has been described in detail by SARFERT *et al.* (1979). In parallel, the morphologically different phage TaZ was used (DNA content $37 \cdot 10^6$ daltons/phage).

Media and experimental conditions: Methods for production, storage and activation of *T. vulgaris* spores, as well as for growth were described by STROHBACH and KRETSCHMER (1977). The interaction with phages and 1227 DNA was studied during growth and differentiation on modified CZAPEK-DOX(CD)-agar, consisting of sucrose 1%, casaminoacids vitamin free (DIFCO) 0.15%, casein peptone SERVA 0.15%, NaCl 0.1%, KH_2PO_4 0.12%, $Na_2HPO_4 \cdot 2\,H_2O$ 0.28%, $CaCO_3$ 0.0005%, $FeSO_4 \cdot 7\,H_2O$ 0.001%, $MgSO_4$ 0.05%, agar 2%, pH 6.3. Incubation occurred at 52 °C. Phages, resp. 1227 DNA were spread onto the mycelial lawn at hourly intervals during the 15 hours of development. Transformation of 1261 mycelium by 1227 DNA was limited by adding DNase (5 µg/plate) after one hour. After about 40 h the spores were harvested separately from all plates and tested for the frequency of transformation of the *nic*-marker in the case of 1227 DNA treatment. The frequency of trapping in the phage treated samples was determined after heating for 15 min at 90 °C to destroy free phages. Then the number of p.f.u. indicated the number of carrier spores. In the latter series also the degree of lysis as a consequence of phage production was determined macroscopically.

The transforming DNA of strain 1227 was prepared by E. SARFERT according to MARMUR (1961). Production of phage lysates (titre about $1:10^{10}$ phages/ml) was described by SARFERT *et al.* (1979).

The morphological development of the mycelia was followed by taking photographs of material removed at hourly intervals from control plates (phase contrast).

Results

Since the study was initiated in order to explain the unexpected results of our transfection experiments (KRETSCHMER and SARFERT 1980), we delt only with competent strains of *T. vulgaris* developing under conditions favourable for expression of competence. The transformation method used (HOPWOOD and WRIGHT 1972) included growth and subsequent sporulation on CD-agar. Despite formation of aerial mycelium by the strains 1227 and 1261 at about the 7th h of incubation, sporulation occurred only within the substrate mycelium. Microscopically the aerial hyphae could be distinguished from substrate hyphae by a thicker diameter, lack of branching and the tendency to break into rods. Using agar of another composition the aerial mycelium was found to sporulate well. Recently, we found the aerial hyphae to differ from substrate mycelium in many respects besides morphology. The behaviour of the sporulating aerial mycelium will be described separately.

Studies with liquid cultures were not included, since there the competent strains did not sporulate. Other strains sporulating well in liquid medium upon exhaustion of the nitrogen source were found to lack competence and to trap phage DNA poorly.

Morphology of the life cycle of substrate mycelium

During the development on CD-agar a specific sequence of morphological changes was observed. Typical stages are represented in Fig. 1. Despite the lack of absolute synchrony caused by the mycelial growth pattern, the sequential order and duration of 6 main stages was determined (Table 1). The secondary mycelium appearing after 7.5 h of incubation differed from the primary mycelium by a high degree of branching and its ability to form endospores after it had stopped growth. But by no means does growing secondary mycelium represent an early sporulation stage; sometimes it fails to start sporulation at all. Commitment to sporulation was only observed at the stage of formation of emergences which appeared about 3.5 h after initiation of secondary growth.

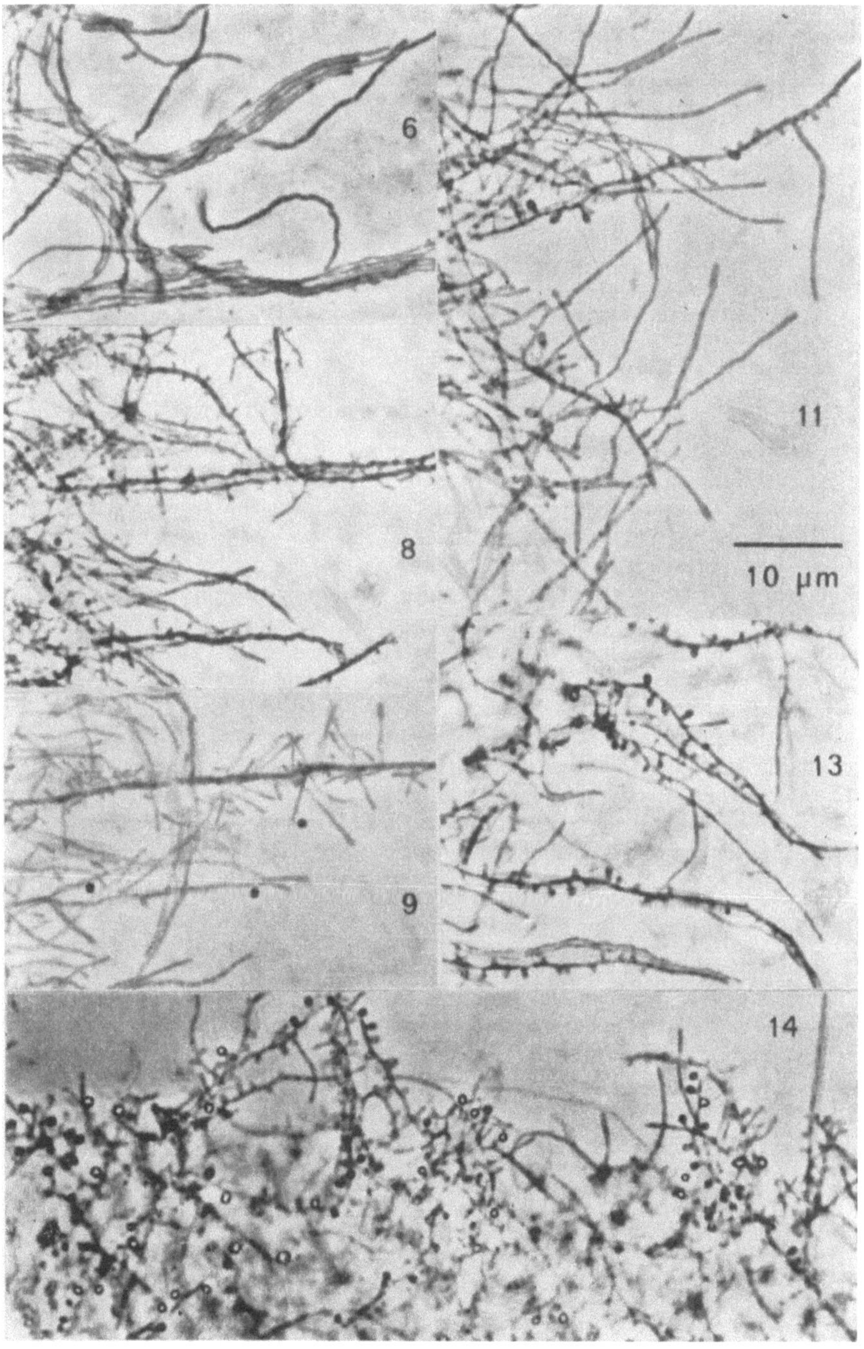

Fig. 1. Main developmental stages of *T. vulgaris* 1261 *amy⁻* substrate mycelium during 6.—14. h of incubation on CD-agar. Phase contrast microscopy.

6: primary mycelium, 8 and 9: formation of secondary mycelium, 11: septated secondary mycelium, 13: secondary mycelium with emergences in which forespores develop, 14: forespores mostly maturated to endospores

Table 1
Sequence of morphologically different stages of the life cycle of *T. vulgaris* 1261 *amy*⁻ substrate mycelium growing on CD-agar

Stage	Developmental period	Duration (h)
I	germination	0 – 0.8
II	primary growth	0.8– 6.5
	lag of growth (after exhaustion of main nutrients, KRETSCHMER 1978a)	6.5– 7.5
III	secondary growth	7.5–10.0
IV	septation of secondary mycelium	10.0–11.0
V	formation of emergences in which forespores are formed	11.0–12.5
VI	maturation of forespores to endospores	12.5–15.0

Modes of interaction of substrate mycelium with phages

Preliminary studies revealed that upon phage Ta1 or TaZ infection young *T. vulgaris* mycelium underwent lysis caused by propagation of phages. However, older mycelium attained resistance against phages and the subsequent sporulation went on unaffectedly. The appearance of resistance was not caused by a loss of phage receptors, since entrance of the phage genome could be proven by the phenomenon of phage trapping: phages added to the late secondary mycelium gave rise to Ta1- resp. TaZ carrier spores in which the phage genome was locked in a heat stable state. Depending on the multiplicity of infection and the time of phage addition, up to 50% of the harvested spores were carrier spores. Upon outgrowth of carrier spores progeny phages were released.

Studying the interaction of germinating spores with phage Ta1, on the one hand carrier spores were used, and on the other spores which had been infected at the time of inoculation. In Fig. 2 the release of progeny phages Ta1 upon germination is presented. With both systems burst size and kinetics of release were identical. About 40 Ta1 phages per germling are released by the 80th min of incubation.

Correlation between morphological stage and mode of interaction with phage Ta1 or 1227 DNA

The changing interactions with phages and with 1227 DNA in dependence on the morphological development of the substrate mycelium were examined simultaneously. Although there is a certain degree of asynchrony of mycelial development, Fig. 3 shows that each mode of interaction is related to a distinct morphological stage. Independent of the strain used (1227, 1261, 1261 *amy*⁻), the primary mycelium was the only stage which propagated and released phage Ta1. Infection at later times left the mycelial development unaffected. During the period of secondary growth phages were lost while transforming DNA was taken up and integrated into the 1261 genome. At the end of the secondary growth phase when the frequency of crosswall formation increased and emergences appeared, competence was lost, but infection with phage led to phage trapping by the developing spores. Later, about 2.5 h before the spores became refractile, the mycelium no longer responded to phage addition.

Discussion

Morphology of the life cycle of substrate mycelium

Using conditions favourable for expression of competence, the competent strains developed in the same way as strain G4 growing as a microculture under a coverglass (KRETSCHMER 1978a, 1978b). The slowly growing secondary mycelium is characterized

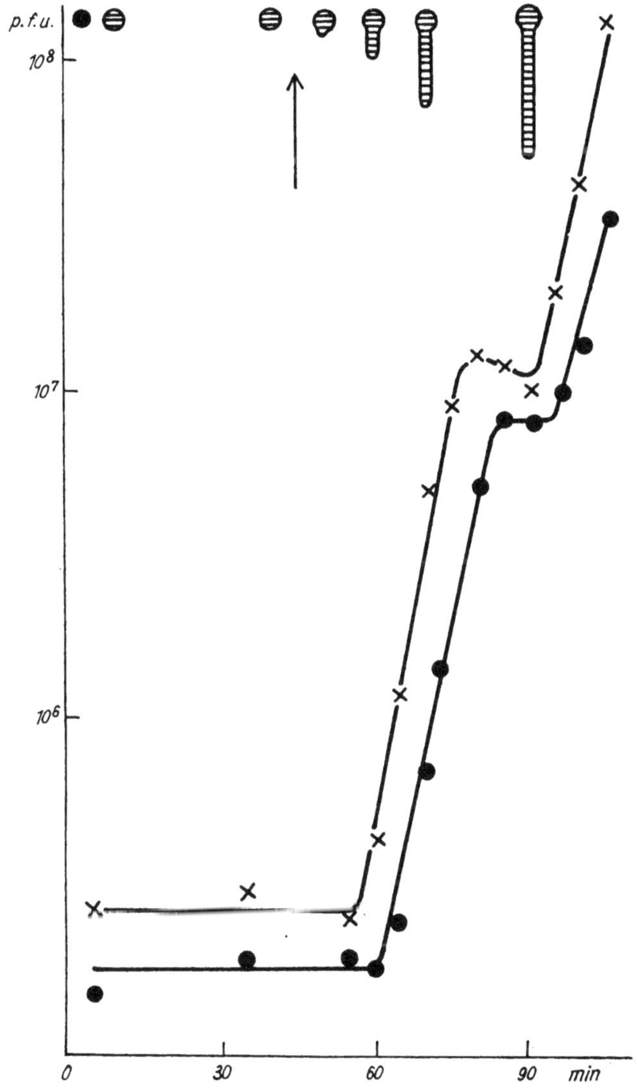

Fig. 2. Release of phage Ta1 by outgrowing 1227 spores.
×: spores infected at the time of inoculation (MOI = 1), ●: spores harboring Ta1 DNA, : ↑ start of outgrowth

by a high degree of branching (about 10 times higher than in primary mycelium) and the ability to sporulate after cessation of growth (Fig. 1). Apparently also in the strains 1227 and 1261 cessation of primary growth and the subsequent initiation of secondary growth is due to a nutritional shift-down.

Interaction of different developmental stages with phages

While the virulent phages Ta1 and TaZ propagate within the primary mycelium, their fate in the secondary mycelium is obscure. The lack of phage production cannot be attributed to a failure of phage adsorption to secondary mycelium, since at the

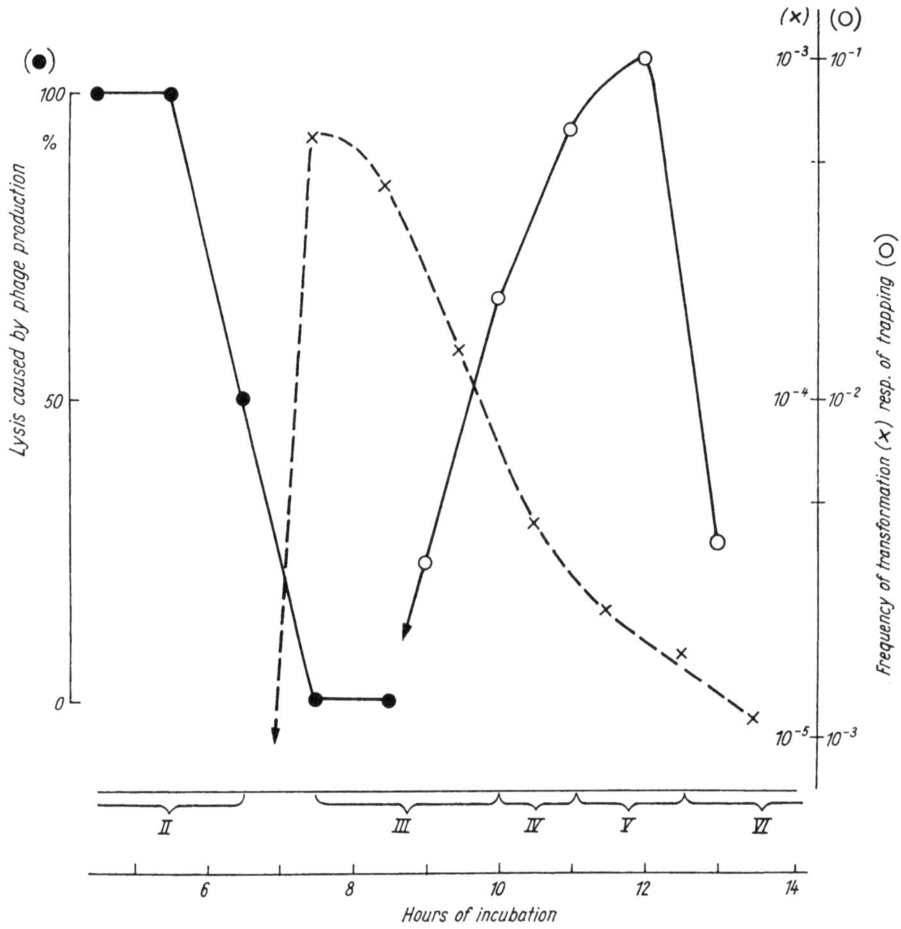

Fig. 3. Changes of frequencies of phage propagation (expressed by mycelial lysis), prophage trapping and transformation in the course of development of *T. vulgaris* 1261 *amy*⁻. Onto mycelia growing on CD-agar from 5.—13. h at hourly intervals either 0.1 ml of a Ta1 suspension (MOI = = 5/spore) or 25 μg *T. vulgaris* 1227 DNA was spread. The DNA on the plates was destroyed one hour later by spreading 5 μg DNase. Stages I—VI are described morphologically in Tab. 1

stage of competence after addition of phage DNA there is also no phage production (KRETSCHMER and SARFERT 1980), and later on adsorption is documented by phage trapping (Fig. 3). We rather assume that the inability to produce phages is caused by an alteration of gene expression in the secondary mycelium. Lack of phage propagation has been observed in sporulating bacilli (GOLDBERG and GOLLAKOTA 1961, YEHLE and DOI 1967, SONENSHEIN and ROSCOE 1969, KAWAMURA and ITO 1975, MORENO 1977). It could be correlated to changes of sigma factors associated with the RNA polymerase (YOUNG 1978, SONENSHEIN and CAMPBELL 1978). The sigma factors of the RNA polymerase determine its template specificity during transcription. The selective recognition of a promotor of a cloned early sporulation gene of *Bacillus subtilis* by altered sigma factors has been described recently (HALDENWANG and LOSICK 1980, HALDENWANG *et al.* 1981, MORGAN *et al.* 1981). However, in addition ribosomal function changes may be necessary to translate spore-specific messages

(SONENSHEIN and CAMPBELL 1978). In any case, in sporulating bacilli the sporulation-associated template specificity change of the host RNA polymerase is sufficient to explain its failure to transcribe the early phage genes (LOSICK and PERO 1981).

Probably the inability of the secondary mycelium to propagate phages is caused by similar transcriptional changes. But, although *T. vulgaris* forms endospores in the same way as bacilli (DOROKHOVA *et al.* 1970, CROSS *et al.* 1971), the existence of an already differentiated growing stage in *T. vulgaris* is unique among the procaryotes. The lack of phage propagation in bacilli is restricted to the sporulation process, since infection of mutants defective for sporulation is always productive, also during the stationary growth phase (SONENSHEIN and ROSCOE 1969, MORENO 1977). The existence of an individually differentiated stage separate from primary and sporulating mycelium is documented further by the ability of the growing secondary mycelium to produce a heat-stable alkaline serine protease (KRETSCHMER *et al.* 1982).

As to the fate of the injected phage DNA within the secondary mycelium, three periods can be distinguished. 1. Adding phages during growth of the secondary mycelium leads to a loss of phages upon further development. This loss is not due to degradation of the phage DNA, since phages are produced if the mycelium is forced to resume vegetative growth using the improved transfection method (KRETSCHMER and SARFERT 1980). Therefore we assume that the phage DNA remains unexpressed within the mycelium and is degraded during lysis of the spore mother cell after completion of the endospore. 2. If phages are added later after cessation of secondary growth and initiation of sporulation, for the following 2.5 h phage DNA becomes locked into the future spore. Of course, the trapping phenomenon indicates that at this time infection occurs in such a way that the phage DNA becomes associated with structures of the future spore. The formation of carrier spores upon phage trapping is also known from *Bacillus subtilis* and *Bacillus cereus*. It takes place if virulent phages are added to cultures at early stages of sporulation (GOLDBERG and GOLLAKOTA 1961, TAKAHASHI 1964, YEHLE and DOI 1967). 3. Infection of the secondary mycelium at later stages of sporulation again results in a loss of the phages. The end of the period, where trapping is possible, apparently coincides with the completion of the forespore. At this time the spore protoplast is surrounded by a double layer of cytoplasmic membrane and of cytoplasm of the spore mother cell, thus being out of reach of infecting phages.

Concerning the ability of germinating spores to propagate phages, spores harboring phage DNA may be suitable objects, since normally the spore wall prevents infection. Indeed, considering that the latency period of Ta1 development is 15 min (SARFERT *et al.* 1979) spores are infected only at the time of outgrowth *i.e.* disruption of the spore wall (Fig. 2). However, the similar kinetics of phage release from infected and from carrier spore germlings indicate that in any case phage development starts only at the moment of outgrowth. Despite RNA synthesis within the spores started immediately after inoculation, the synthesis of DNA commences only at the time of outgrowth (STROHBACH and KRETSCHMER 1977). Thus, the behaviour of the carrier spores indicates that upon germination the vegetative state is restored only 45 min after inoculation. Similar results were reported for *B. cereus* spores carrying CP-51 DNA and *B. subtilis* spores harboring phage Φe DNA (COHEN *et al.* 1973, BUU and SONENSHEIN 1975). The delay of initiation of phage development was correlated to transcriptional peculiarities of the spore RNA polymerase.

Correlation of expression of competence to developmental stages

Comparing the appearance of competence in *B. subtilis* and *T. vulgaris*, in both organisms the competent state is expressed at a stage intermediate between vegetative growth and sporulation. However, the competent stage represents different developmental states in these organisms, which is also obvious from the different fates of the incorporated DNA. In *T. vulgaris* DNA is taken up by the growing secondary mycelium which allows integration of the 1227 DNA into the future spore genome, but excludes phage DNA from direct propagation or trapping. In contrast, competent *B. subtilis* cells propagate phages upon treatment with phage DNA (GREEN 1964, REILLY and SPIZIZEN 1965, KRETSCHMER and SARFERT 1971) while *B. subtilis* DNA is not integrated into the future spore genome (KRETSCHMER 1968, 1969a, 1973). Thus interruption of the differentiation process and resumption of vegetative growth is necessary for transformation of *B. subtilis*, but for transfection in *T. vulgaris* (KRETSCHMER and SARFERT 1980). In *B. subtilis* only those cells become competent which for endogenous reasons stop growing in an unlimited growing culture. After cessation of growth these cells start sporulation 150 min later. During the last 50 min of this praesporulation period competence appears (KRETSCHMER 1969b, 1973).

Since in *T. vulgaris* trapping does not occur at the competent stage (Fig. 3), the appearance of carrier spores besides free phages in the improved transfection experiment must have been caused by indirect infection of the finally sporulating mycelium with the phages released after transfection.

In summary, studying the interaction with DNA (Tal or 1227) or phages has proved to be a suitable way to characterize developmental stages of *T. vulgaris* beyond morphological description.

Acknowledgement

The skillful technical assistance of Mrs. GERDA ELSKE is gratefully acknowledged.

References

BUU, A. and SONENSHEIN, A. L., 1975. Nucleic acid synthesis and ribonucleic acid polymerase specificity in germinating and outgrowing spores of *Bacillus subtilis*. J. Bacteriol., **124**, 190—200.

COHEN, A., BEN-ZE'EV, H. and YASHOUV, J., 1973. Outgrowth of *Bacillus cereus* spores harboring bacteriophage CP-51 DNA. I. Initiation of bacteriophage development. J. Virol., **11**, 648—654.

CROSS, T., DAVIES, F. L. and WALKER, P. D., 1971. *Thermoactinomyces vulgaris*. I. Fine structure of the developing endospores. In: Spore Research 1971, p. 175 (edited by BARKER, A. N., GOULD, G. W. and WOLF, J.). Academic Press London—New York.

DOROKHOVA, L. A., AGRE, N. S., KALAKOUTSKII, L. V. and KRASSILNIKOV, N. A., 1970. Electron-microscopic study on spore formation in *Micromonospora vulgaris*. Mikrobiologiya, **39**, 680—684.

GOLDBERG, I. D. and GOLLAKOTA, K., 1961. Bacteriophage in *Bacillus cereus*. In: Spores II. (ed. HALVORSON, H. O.). Burgess Publ. Comp., USA, p. 276—296.

GREEN, D. M., 1964. Infectivity of DNA isolated from *Bacillus subtilis* bacteriophage SP 82. J. molecular Biol., **10**, 438—451.

HALDENWANG, W. G. and LOSICK, R.,1980. A novel RNA polymerase sigma factor from *Bacillus subtilis*. Proc. Nat. Acad. Sci. USA, **77**, 7000—7004.

HALDENWANG, W. G., LANG, N. and LOSICK, R., 1981. A sporulation-induced-sigma-like regulatory protein from *Bacillus subtilis*. Cell, **23**, 615—624.

HOPWOOD, D. A. and WRIGHT, H. M., 1972. Transformation in *Thermoactinomyces vulgaris*. J. gen. Microbiol., **71**, 383—398.

KAWAMURA, F. and ITO, J., 1975. Bacteriophage gene expression in differentiating cells of *Bacillus subtilis*. In: Spores VI (edited by GERHARD, P., COSTILOW, R. N. and SADOFF, H. L.). American Soc. for Microbiology, Washington, D. C., p. 231—240.

KRETSCHMER, S., 1968. Stabilisierung potentieller Transformanten von *Bacillus subtilis*. Z. Allg. Mikrobiol., **8**, 155—159.

KRETSCHMER, S., 1969a. Stabilität potentieller Transformanten bei Sporulationsmutanten von *Bacillus subtilis*. Z. Allg. Mikrobiol., **9**, 333—336.

KRETSCHMER, S., 1969b. Kompetenzdauer von *Bacillus subtilis*-Zellen. Z. Allg. Mikrobiol., **9**, 457–470.

KRETSCHMER, S., 1973. Sporenbildung in wachsenden *Bacillus subtilis*-Kulturen und ihre Korrelation zur Kompetenz für genetische Transformation. Z. Allg. Mikrobiol., **12**, 221–232.

KRETSCHMER, S., 1978a Transition of *Thermoactinomyces* substrate mycelium from growth to sporulation. Z. Allg. Mikrobiol., **18**, 613–616.

KRETSCHMER, S., 1978b. Kinetics of vegetative growth of *Thermoactinomyces vulgaris*. Z. Allg. Mikrobiol., **18**, 701–711.

KRETSCHMER, S. and SARFERT, E., 1971. Vergleich potentieller Transformanten und Transfektanten von *Bacillus subtilis*. Z. Allg. Mikrobiol., **11**, 633–636.

KRETSCHMER, S. and SARFERT, E., 1980. Transfection in *Thermoactinomyces vulgaris*. Z. Allg. Mikrobiol., **20**, 73–75.

KRETSCHMER, S., KÖRNER, D., STROHBACH, G., KLINGENBERG, P., JAKOB, H.-E., GUMPERT, J. und RUTTLOFF, H., 1982. Physiologische und zellbiologische Charakterisierung des Proteasebildners *Thermoactinomyces vulgaris* während langzeitiger Kultivierung im Rührfermentor. Z. Allg. Mikrobiol., **22**, 693–703.

LOSICK, R. and PERO, J., 1981. Cascades of sigma factors. Cell, **25**, 582–585.

MARMUR, J., 1961. A procedure for the isolation of deoxyribonucleic acid from microorganisms. J. molecular Biol., **3**, 208–218.

MORENO, F., 1977. Development of bacteriophage *Φ* 29 in sporulating and non-sporulating cells of *Bacillus subtilis* 168. Annal. Microbiol., **128**, 3–18.

MORGAN, C. P., LANG, N., BANNER, C. D. B., HALDENWANG, W. G. and LOSICK, R., 1981. Promoter for a developmentally regulated gene in *Bacillus subtilis*. Cell, **25**, 783–793.

REILLY, B. E. and SPIZIZEN, J., 1965. Bacteriophage deoxyribonucleate infection of competent *Bacillus subtilis*. J. Bacteriol., **89**, 782–790.

SARFERT, E., KRETSCHMER, S., TRIEBEL, H. and LUCK, G., 1979. Properties of *Thermoactinomyces vulgaris* phage Ta1 and its extracted DNA. Z. Allg. Mikrobiol., **19**, 203–210.

SONENSHEIN, A. L. and ROSCOE, D. H., 1969. The course of phage *Φ* infection in sporulating cells of *Bacillus subtilis* strain 3610. Virology, **39**, 265–276.

SONENSHEIN, A. L. and CAMPBELL, K. M., 1978. Control of gene expression during sporulation. In: Spores VII (Eds. CHAMBLISS, G. and VARY, J. C.), p. 179–192. American Soc. Microbiol., Washington DC.

STROHBACH, G. und KRETSCHMER, S., 1977. Einbau von Thymidin in die DNS von Actinomyceten. I. Der Einbau exogenen Thymidins in die DNS von *Thermoactinomyces vulgaris*. Z. Allg. Mikrobiol., **17**, 559–568.

TAKAHASHI, I., 1964. Incorporation of bacteriophage genome by spores of *Bacillus subtilis*. J. Bacteriol., **87**, 1499–1502.

YEHLE, C. O. and DOI, R. H., 1967. Differential expression of bacteriophage genomes in vegetative and sporulating cells of *Bacillus subtilis*. J. Virol., **1**, 935–947.

YOUNG, M., 1978. Bacterial endospore development — an ordered sequence of gene transcription. TIBS, March 1978, 55–59.

Mailing address: Dr. S. KRETSCHMER
Zentralinstitut für Mikrobiologie
und experimentelle Therapie der AdW
DDR 6900 Jena, Beutenbergstr. 11

Zeitschrift für Allgemeine Mikrobiologie	22	9	1982	639–647

(Akademie der Wissenschaften der DDR,
Forschungszentrum für Molekularbiologie und Medizin,
Zentralinstitut für Mikrobiologie und experimentelle Therapie, Jena,
Direktor: Prof. Dr. habil. U. TAUBENECK)

Genetische Rekombination in Mutanten des Anthracyclin-Bildners *Streptomyces griseus*

G. SCHUMANN und W. F. FLECK

(Eingegangen am 7. 4. 1982)

It was found that genetic recombination occurs if two marked strains of *Streptomyces griseus* (leukaemomycin-producing strains IMET JA 3933 and IMET JA 5142) are grown together in mixed cultures on semisolid media. The crossing techniques used and the method for carrying out selective analysis were essentially the same as those described by HOPWOOD (1967, 1972). The parent strains used for crosses were marked with single or double nutritional requirements and with mutations for drug resistance. The crosses are quite self-sterile, yielding only in one combination stable prototrophic recombinants at a low frequency (10^{-5} to 10^{-6}). The majority of recombinants behaved as stable haploid genotypes. A series of four-point crosses of different types of auxotrophs was carried out. The results of these experiments do not provide sufficient data for constructing a chromosome map, but provide basic information on the possibilities of genetic analysis of the production of anthracycline antibiotics. The majority of crosses performed were not fertile at 28 °C but, surprisingly, in some crosses carried out at 34 °C viable colonies were detected on minimal media at frequencies from 10^{-3} to 10^{-2}.

Genetische Rekombination bei Antibiotica-bildenden Streptomyceten ist mehrfach nachgewiesen (SERMONTI u. SPADA-SERMONTI 1955; Übersicht: HOPWOOD u. MERRICK 1977). Die bisher am besten untersuchten Rekombinationssysteme bei Streptomyceten sind das von HOPWOOD beschriebene bei *Streptomyces coelicolor* (1973) und das von *Streptomyces olivaceus* (MATSELYUKH 1976). Bei industriell bedeutsamen Streptomyceten sind Rekombinationssysteme für Bildner der Antibiotica Oxytetracyclin (ALACEVIC 1969, FRIEND u. HOPWOOD 1971), Zorbamycin (COATS u. ROESER 1971), Chloramphenicol (FRANCIS *et al.* 1975), Cephalosporine (KIRBY 1978) und Cephamycin (PARAG 1978) aufgebaut und zur Aufstellung von Chromosomenkarten herangezogen worden. Ein Rekombinationssystem bei Anthracyclin-bildenden Streptomyceten-Arten ist bisher nicht bekannt. Die vorliegende Arbeit hatte daher das Ziel, ein Rekombinationssystem bei Leukaemomycin-bildenden *Streptomyces griseus*-Stämmen aufzubauen und durch genetische Analysen Voraussetzungen für die Genkartierung zu schaffen.

Material und Methoden

Verwendete Stämme: Als Ausgangsstämme zur Gewinnung auxotropher Mutanten sind Wildstämme des Leukaemomycin-Bildners *Streptomyces griseus* (IMET JA 5142 und IMET JA 3933) eingesetzt worden. Leukaemomycin ist ein Komplex von Anthracyclin-Antibiotica (FLECK u. STRAUSS 1975), der eine mit dem klinisch angewandten Daunorubicin identische Komponente enthält.

Medien: Zur Stammerhaltung, Mutantenisolierung, zum Anlegen von Mischkulturen und zur Bestimmung des Antibioticum-Bildungsvermögens eignete sich ein Hafermehl-Agar (HMA) folgender Zusammensetzung (g/l): Hafermehl 20, Agar 15 (pH 7,5). Als Minimal-Medium (MM) diente das von HOPWOOD (1967) beschriebene. Supplemente zum MM wurden als sterile Lösungen zugesetzt. Bei der Charakterisierung auxotropher Mutanten bzw. Rekombinanten sind folgende Endkonzentrationen zugesetzter Metabolite erreicht worden: Aminosäuren 50 mg/l, Purine und Pyrimidine 10 mg/l, Vitamine 1 mg/l. Zum Vorkeimen der Sporen vor der Mutagenbehandlung ist Medium 79 (Dextrose 10, Bacto-Pepton/DIFCO 10, Casaminoacids/DIFCO 1, Hefeextrakt/DIFCO 2, NaCl 6 g/l) verwendet worden.

Mutagenbehandlung: Die für Rekombinationsexperimente eingesetzten Mutanten wurden mit Hilfe des N-Methyl-N'-nitro-N-nitrosoguanidin (MNNG) nach der von DELIC *et al.* (1970) beschriebenen Methode erzeugt und selektiert. Die Herstellung von Suspensionen mit Einzelsporen erfolgte durch Abschwemmen 14 Tage alter Emerskulturen auf HMA, anschließender Sporentrennung mit Hilfe des Vibrationshomogenisators VHG1 und nachfolgender Filtration durch G3-Glasfritten (SCHOTT & Gen.). Nach Zentrifugieren der Einzelsporen-Suspension sind die sedimentierten Sporen in 0,05 M TRIS-Maleinsäure-Puffer (pH 9,0) resuspendiert worden. In dieser Suspension wurden 1 mg MNNG/ml (ALDRICH Chemical Company/Milwaukee/Wisconsin — USA) gelöst und 60 min bei 37 °C inkubiert. Um die Mutantenausbeute zu erhöhen, sind die Sporen vor der Mutagenbehandlung 60 oder 90 min bei 28 °C in Medium 79 vorinkubiert worden. Die vorgekeimten Sporen wurden unter den o. a. Bedingungen nur 30 min mit MNNG inkubiert. Bei den gewählten Inkubationszeiten ergaben sich Überlebensraten von 0,01 bis 1%.

Kreuzung und genetische Analyse: Zur Kreuzung von Mutanten sowie zur Selektivanalyse der Rekombinanten sind die von HOPWOOD (1967 und 1969) beschriebenen Methoden eingesetzt worden. Kreuzungen wurden durch Mischen der Sporen zweier Mutanten-Stämme mit komplementärem Genotyp auf HMA-Platten ausgeführt. Mischkulturen wurden (a) 14 Tage bei 28 °C oder (b) 4 Tage bei 34 °C und anschließend 10 Tage bei 28 °C inkubiert. Sporen der Mischkulturen wurden mit Hilfe der Replica-Technik von LEDERBERG u. LEDERBERG (1952) auf Selektiv-Medien übertragen. Wenn auf diese Weise Rekombinanten beobachtet wurden, sind die Sporen rekombinationsfähiger Mischkulturen in isotonischer NaCl-Lösung suspendiert und in geeigneten Verdünnungen auf Selektiv-Medien ausgeplattet worden. Die auf Selektiv-Medien wachsenden Rekombinanten wurden auf „master plates" mit dem gleichen Medium überimpft. Nach Versporung der Rekombinanten wurden diese zur Bestimmung ihres Phänotyps auf Differenzierungsmedien überstempelt.

Ergebnisse

Mutagenbehandlung

MNNG-Behandlung von Sporen der Wildstämme IMET JA 5142 und IMET JA 3933 ohne Präinkubation führte zu Überlebensraten von 0,1 bis 1%. Bei etwa 0,1% der Überlebenden handelte es sich um stabil-auxotrophe Mutanten. Die Mehrzahl der Mutanten war Cystein-auxotroph oder in ihrer Auxotrophie nicht zu bestimmen.

MNNG-Behandlung nach Präinkubation der Sporen führte demgegenüber zu Überlebensraten von 0,001 bis 0,1%. Bis zu 3% der Überlebenden waren stabil-auxotrophe Mutanten.

In Tabelle 1 sind Mutanten aufgeführt, die auf Grund guter Versporung und geringer Reversionsrate ($<10^{-7}$) günstige Voraussetzungen für Kreuzungsexperimente mitbringen. Um doppeltauxotrophe Mutanten als Kreuzungspartner zu erhalten, sind ausgewählte Auxotrophe einer nochmaligen MNNG-Behandlung unterzogen worden. Die für Rekombinationsversuche eingesetzten doppeltauxotrophen Mutanten sind in Tabelle 2 aufgeführt.

Kreuzungsexperimente

Alle möglichen Kombinationen der in Tabelle 1 aufgeführten einfach-auxotrophen Mutanten wurden in Kreuzungsversuchen auf Rekombinantenertrag geprüft. Dabei sind die Mischkulturen unter optimalen Wachstumsbedingungen bei 28 °C inkubiert worden. Stabile prototrophe Rekombinanten traten lediglich in der Kombination

IMET JA 5142-01 *cys1* × IMET JA 3933-3 *aro1 lkmH*

mit einer Rate von 10^{-6} bis 10^{-5} auf. Alle Rekombinanten trugen den Marker *lkmH*.

Tabelle 1
Für Kreuzungsexperimente eingesetzte einfachauxotrophe Mutanten des Leukaemomycin-Bildners *Streptomyces griseus*

Ausgangsstamm	Marker	Anzahl isolierter Mutanten	Auxotrophie
IMET JA 5142	*cys*	5	Cystein
	met	2	Methionin
	his	2	Histidin
	ser	1	Serin
IMET JA 3933	*cys*	6	Cystein
	ade	3	Adenin
	aro	1	Aromatische Aminosäuren[1])
	ura	1	Uracil
	ser	1	Serin
	arg	1	Arginin
	lys	1	Lysin
		7	nicht bestimmt

[1]) Tyrosin, Tryptophan, Phenylalanin

Tabelle 2
Für Kreuzungsexperimente eingesetzte doppeltauxotrophe Mutanten des Leukaemomycin-Bildners *Streptomyces griseus*[1])

Mutanten-Bezeichnung	Marker	Auxotrophie
IMET JA 5142-01/6	*cys1 ade4*	Cystein, Adenin
IMET JA 5142-01/41	*cys1 aro2*	Cystein, aromatische AS
IMET JA 5142-01/14	*cys1 thi1*	Cystein, Thiamin
IMET JA 5142-01/27	*cys1 met3*	Cystein, Methionin
IMET JA 3933-3/27	*aro1 ser3 lkmH*[2])	Aromatische AS, Serin
IMET JA 3933-31/5	*aro1 his3 lkmH*	Aromatische AS, Histidin
IMET JA 3933-3/30	*aro1 met4 lkmH*	Aromatische AS, Methionin
IMET JA 3933-31/17	*aro1 cys12 lkmH*	Aromatische AS, Cystein
IMET JA 3933-31/15	*aro1 ade5 lkmH*	Aromatische AS, Adenin
IMET JA 3933-31/12	*aro1 pdx1 lkmH*	Aromatische AS, Pyridoxin
IMET JA 3933-3/39	*aro1 ura2 lkmH*	Aromatische AS, Uracil
IMET JA 3933-31/3	*aro1 ilv1 lkmH*	Aromatische AS, Isoleucin + Valin
IMET JA 3933-11/13	*arg1 met5*	Arginin, Methionin
IMET JA 3933-11/3	*arg1 ade6*	Arginin, Adenin
IMET JA 3933-11/6	*arg1 ilv2*	Arginin, Isoleucin + Valin
IMET JA 3933-11/24	*arg1 leu1*	Arginin, Leucin
IMET JA 3933-11/8	*arg1 ura3*	Arginin, Uracil
IMET JA 3933-11/12	*arg1 cys13*	Arginin, Cystein
IMET JA 3933-27/3	*lys1 cys14*	Lysin, Cystein
IMET JA 3933-27/4	*lys1 arg2*	Lysin, Arginin
IMET JA 3933-27/1	*lys1 ade7*	Lysin, Adenin
IMET JA 3933-32/5	*ura1 arg3 lkmH*	Uracil, Arginin
IMET JA 3933-2/2	*ser2 met6*	Serin, Methionin
IMET JA 3933-24/7	*pdx3 mth1*	Pyridoxin, Methionin + Threonin

[1]) Außerdem ist die dreifachauxotrophe Mutante IMET JA 3933-27/5 mit den Markern *lys1 aro3 pdx2* (auxotroph für Lysin, aromatische AS und Pyridoxin) eingesetzt worden
[2]) *lkm* = genetischer Block in der Leukaemomycin-Biosynthese, *lkmH* = genetischer Block in der Leukaemomycin-Biosynthese, der zur Bildung von epsilon-Rhodomycinon führt (Wagner *et al.* 1981)

Die aus diesen Mutanten gewonnenen doppeltauxotrophen Mutanten (Tab. 2) sind ebenfalls zu Kreuzungsexperimenten herangezogen worden. Allerdings konnten in keinem Falle Rekombinanten nachgewiesen werden.

Nach Variation der Kulturbedingungen wurde überraschenderweise gefunden, daß Inkubation der Mischkulturen bei 34 °C anstatt 28 °C bei einem Teil der Vierpunkt-Kreuzungen zu Rekombinationen führt. Bezogen auf die Gesamtnachkommenschaft der Mischkultur wurden dabei Rekombinationsraten bis zu 10^{-2} erreicht.

Da die verwendeten Kreuzungspartner bei 34 °C kaum Luftmycel ausbilden können, wurden die Mischkulturen nach der Inkubation bei 34 °C anschließend mehrere Tage bei 28 °C weiterbebrütet, ehe ausreichend Sporen ausdifferenziert waren.

Tabelle 3
Rekombinationsraten in 2 unterschiedlichen Vierpunkt-Kreuzungen mit Mutanten des Leukaemomycin-Bildners *Streptomyces griseus* in Abhängigkeit von der Inkubationsdauer der Mischkulturen bei 34 °C

Inkubationsdauer (Tage)/ Temperatur	Rekombinationsrate[1]) Kreuzung 1	Kreuzung 2
14/28 °C	$<1 \times 10^{-7}$	$<5 \times 10^{-6}$
1/34 °C + 13/28 °C	$<1 \times 10^{-7}$	5×10^{-7}
2/34 °C + 12/28 °C	1×10^{-5}	1×10^{-3}
4/34 °C + 10/28 °C	3×10^{-3}	2×10^{-3}
7/34 °C + 7/28 °C	1×10^{-2}	2×10^{-3}

[1]) Kreuzung 1: IMET JA 5142-01/6 *cys1 ade4* × IMET JA 3933-3/39 *aro1 ura2 lkmH*
Kreuzung 2: IMET JA 5142-01/6 *cys1 ade4* × IMET JA 3933-3/27 *aro1 ser3 lkmH*

In einer Versuchsreihe wurde die optimale Inkubationsdauer bei 34 °C ermittelt. Die Ergebnisse sind in Tabelle 3 zusammengefaßt. Sie zeigen, daß die Inkubationsdauer bei 34 °C mindestens 2 Tage und maximal 4 Tage betragen muß, wenn die in Tabelle 3 aufgeführten Rekombinationsraten erreicht werden sollen.

Auf Grund dieser Ergebnisse sind alle weiteren Kreuzungsexperimente so durchgeführt worden, daß die Mischkulturen zunächst 4 Tage bei 34 °C und anschließend 10 Tage bei 28 °C inkubiert wurden.

Die in Tabelle 2 ersichtlichen Mutanten mit Doppelauxotrophie sind in allen möglichen Kombinationen bei 34 °C auf Rekombinationsfähigkeit geprüft worden. Dabei zeigte sich, daß lediglich dann Rekombinanten auftraten, wenn als Kreuzungspartner die Mutante IMET JA 5142-01/6 mit dem Genotyp *cys1 ade4* und doppeltauxotrophe Mutanten des IMET JA 3933 (Ausgangsmutanten: *aro1 lkmH* bzw. *arg 1*) eingesetzt wurden.

Die folgenden Kombinationen führten ausschließlich zu stabilen, prototrophen Rekombinanten:

IMET JA 5142-01/6 *cys1 ade4* × IMET JA 3933-3/39 *aro ura lkmH*
IMET JA 5142-01/6 *cys1 ade4* × IMET JA 3933-31/12 *aro pdx lkmH*
IMET JA 5142-01/6 *cys1 ade4* × IMET JA 3933-11/8 *arg1 ura3*

Alle prototrophen Rekombinanten wurden auf Biosynthesevermögen für Leukaemomycine (Anthracyclin-Antibiotica) geprüft. Dabei zeigte sich, daß *ca.* 1000 Rekombinanten aus den ersten beiden Kombinationen in jedem Falle den Marker *lkmH* trugen.

Die Kreuzung IMET JA 5142-01/6 *cys1 ade4* × IMET JA 3933-31/5 *aro his lkmH* führte zu 90% prototrophen Rekombinanten, die ebenfalls alle in der Leukaemomycin-Bildung geblockt waren. Die übrigen 10% der Rekombinanten waren Cystein-auxotroph. Unter den Cystein-auxotrophen Mutanten trugen nur etwa 60% den Marker *lkmH*.

Aus der Nachkommenschaft der Kreuzungen
IMET JA 5142-01/6 *cys1 ade4* × IMET JA 3933-3/27 *aro1 ser3 lkmH*
sowie
IMET JA 5142-01/62 *cys1 ade4 rif2* × IMET JA 3933-11/24 *arg1 leu1*
konnten Vertreter unterschiedlicher Rekombinanten-Klassen isoliert werden. Das ermöglichte die Anwendung genetischer Analysen-Methoden, wie sie für Streptomyceten von Hopwood 1967 und 1969 publiziert wurden. Mit Hilfe dieser Methoden ist die relative Anordnung von Markern auf einer ringförmigen Genkarte beim Leukaemomycin-Bildner *Streptomyces griseus* versucht worden. In den Tabellen 4 bis 7 sind Ergebnisse einer Analyse der Kreuzung von IMET JA 5142-01/6 *cys1 ade4* × IMET JA 3933-3/27 *aro1 ser3 lkmH* dargestellt.

Tabelle 4

Ergebnisse einer Kreuzung IMET JA 5142-01/6 *cys1 ade4* × IMET JA 3933-3/27 *aro1 ser3 lkmH*

Rekombinanten/1,6 × 10^5 koloniebildende Einheiten											
			MM+	Aro	Cys	Aro	Ade	Ser	Cys	Ser	Ade
				340		46		338		102	
			geprüft:	110		56		112		56	
Genotyp			davon:		*lkm*		*lkm*		*lkm*		*lkm*
+	+	+	+	22	12	48	27	13	7	16	9
cys	+	+	+	65	16	—		66	16	—	
+	*ade*	+	+	—	8	8	8	—		0	
+	+	*aro*	+	4	0	0		—		—	
+	+	+	*ser*	—		—		29	29	40	40
cys	+	+	*ser*	—		—		4	4	—	
cys	+	*aro*	+	19	0	—		—		—	

berechnet auf Rekombinanten/1,6 × 10^5 koloniebildende E.

Genotyp			MM+	Aro Cys	Aro Ade	Ser Cys	Ser Ade	∅
+	+	+	+	68	39	39	29	44
cys	+	+	+	200	—	199	—	199
+	*ade*	+	+	—	6	—	0	3
+	+	*aro*	+	12	0	—	—	6
+	+	+	*ser*	—	—	87	73	80
cys	+	+	*ser*	—	—	12	—	12
+	*ade*	*aro*	+	—	0	—		0
cys	+	*aro*	+	59	—	—	—	59
+	*ade*	+	*ser*	—	—	—	0	0

Nach dem Kriterium des minimalen Vierfach-Crossovers (Tab. 5) ist die Anordnung der Marker entsprechend Modell 3 den anderen beiden Möglichkeiten vorzuziehen. Die Analyse der Segregation von nichtselektierten Allelen (Tab. 6) bestätigt diese Schlußfolgerung. Bei Selektion auf *ade*$^+$, *aro*$^+$ kann *lkmH* entsprechend dem Allelfrequenz-Gradienten in die Intervalle *ade-aro* bzw. *cys-ade* eingeordnet werden (Tab. 7). Nach dem Kriterium der minimalen Vierfach-Crossover ist hier die Anordnung von *lkmH* zwischen *ade* und *aro* vorzuziehen.

Tabelle 5
Ordnen der Marker nach minimalem Vierfach-Crossover

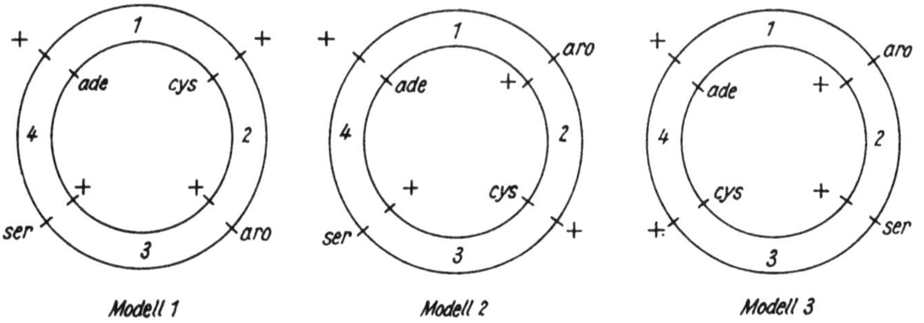

Genotyp				Anzahl	Crossover bei Modell 1	2	3
+	+	+	+	44	2, 4	1, 2, 3, 4	1, 3
cys	+	+	+	199	1, 4	1, 4	1, 4
+	*ade*	+	+	3	1, 2	2, 3	3, 4
+	+	*aro*	+	6	3, 4	3, 4	2, 3
+	+	+	*ser*	80	2, 3	1, 2	1, 2
cys	+	+	*ser*	12	1, 3	1, 3	1, 2, 3, 4
+	*ade*	*aro*	+	0	1, 3	1, 3	1, 2, 3, 4
cys	+	*aro*	+	59	1, 2. 3, 4	2, 4	2, 4
+	*ade*	+	*ser*	0	1, 2, 3, 4	2, 4	2, 4
Anzahl der Vierfach-Crossover:					59	44	12

Tabelle 6
Segregation von nichtselektierten Allelen
Medium 1 (selektierte Allele: ade^+ ser^+)

	aro^+	aro^-	
cys^+	22	4	
cys^-	65	19	$P = 0{,}5-0{,}7$

Medium 3 (selektierte Allele: ade^+ aro^+)

	ser^+	ser^-	
cys^+	13	29	
cys^-	66	4	$P = <0{,}004$

P = Wahrscheinlichkeit der Unabhängigkeit

Nach den gleichen Methoden wurden die Ergebnisse einer Kreuzung IMET JA 5142-01/62 *cys1 ade4 rif1* × IMET JA 3933-11/24 *arg1 leu1* analysiert. Die Rifampicin-resistente Mutante wurde nach Röntgenbestrahlung von Sporen der Mutante IMET JA 5142-01/6 *cys1 ade4* isoliert. Alle Rekombinanten aus dieser Kreuzung waren gegen das Antibioticum Rifampicin sensibel. Bei Selektion auf Rifampicin-Resistenz konnten keine Rekombinanten isoliert werden. Eine Zuordnung der Marker nach dem Kriterium der minimalen Vierfach-Crossover war in dieser Kreu-

Tabelle 7

Einordnen von *lkmH* in Modell 3 nach Allelfrequenz-Gradienten

Medium MM + Ser Cys

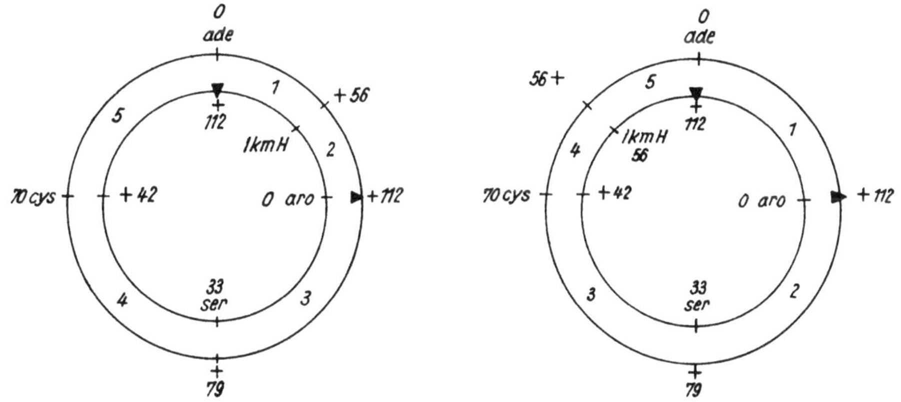

lkmH zwischen					Anzahl	*ade-aro*	*cys-ade*
+	+	+	+	*lkm*	7	2, 4	1, 3
+	+	+	+	+	6	1, 4.	1, 3. 4. 5
cys	+	+	+	*lkm*	16	2, 5	4, 1
cys	+	+	+	+	50	1, 5	5, 1
+	+	+	*ser*	*lkm*	29	2, 3	1, 2
+	+	+	*ser*	+	0	1, 3	1, 2, 4, 5
cys	+	+	*ser*	*lkm*	4	2, 3, 4, 5	1, 2, 3, 5
cys	+	+	*ser*	+	0	1, 3, 4, 5	1, 2, 3, 5
Anzahl des Vierfach-Crossover						4	10

zung nicht möglich. Die Analyse der Segregation nichtselektierter Allele ergibt folgende Anordnung der Marker:

ade4-arg1 — leu1-cys1-ade4

Eine Zuordnung von *arg1 leu1* zu *aro1 ser3* war bisher nicht möglich, weil in den im folgenden aufgeführten Kreuzungen keine Rekombinanten auftraten:

IMET JA 3933-3/27 *aro1 ser3 lkmH* × IMET JA 3933-11/24 *arg1 leu1*

IMET JA 3933-3/27 *aro1 ser3 lkmH* × R53 *cys1 arg1*

IMET JA 3933-3/27 *aro1 ser3 lkmH* × *cys1 leu1*

IMET JA 3933 11/24 *arg1 leu1* × R49 *aro1 cys1*

Bei den als Kreuzungspartner eingesetzten Stämmen R54 und R56 handelt es sich um Rekombinanten aus der Kreuzung 01/62 × 11/24, bei dem Stamm R49 um eine Rekombinante aus der Kreuzung 01/6 × 3/27. Bei der Mehrzahl aller untersuchten Kreuzungen konnte keine genetische Rekombination nachgewiesen werden. Die Nachkommenschaft dieser Kreuzungen zeigte kein Wachstum auf Selektivmedien oder es traten in der Nachkommenschaft instabile Kolonien auf den verwendeten Selektivmedien auf.

Diese Kolonien segregierten in allen untersuchten Fällen ausschließlich beide Eltern-Genotypen im Verhältnis 1:1. Damit handelt es sich um Heterokaryoten, die nach Inkubation der Mischkulturen entstanden. Heteroklone sind dagegen nicht nachgewiesen worden.

Diskussion

Die Ergebnisse zeigen, daß in Mischkulturen von Mutanten der Anthracyclin-bildenden Stämme IMET JA 3933 und IMET JA 5142 der Art *Streptomyces griseus* genetische Rekombination auftritt, wenn die Mischkulturen anstatt bei 28 bei 34 °C inkubiert werden. Dagegen treten in den gleichen Mischkulturen keine Rekombinanten auf, wenn die normale Wachstumstemperatur von 28 °C gewählt wird. Ursachen für das beobachtete Phänomen sind bisher noch nicht bekannt. In den bei 28 °C bebrüteten Mischkulturen ließ sich das Auftreten von Heterokaryose, d. h. der Transfer kompletter Donor-Genome in Rezipienten-Zellen nachweisen. Demgegenüber geht aus der Analyse der bei 34 °C entstandenen Rekombinanten hervor, daß lediglich Teile des Donor-Genoms an der Bildung der Rekombinanten beteiligt sind. Es wird daher angenommen, daß sich bei den untersuchten Mutanten Art und Effektivität des Gentransfers in Abhängigkeit von der Inkubationstemperatur der Mischkulturen unterscheidet. Daß die analysierten Kreuzungen streng polar waren, bestätigt die Annahme, wonach ein Kreuzungspartner stets als Donor fungierte und lediglich ein Genomfragment in den Rezipienten transferierte. Diese Befunde entsprechen den von HOPWOOD *et al.* (1969) bei *Streptomyces coelicolor* gefundenen Verhältnissen nach NF × UF-Kreuzungen. Allerdings ist unter den untersuchten Mutanten lediglich der Genotyp *cys1 ade4* bzw. *cys1 ade4 rif1* (IMET JA 5142-01/6 bzw. IMET JA 5142-01/62) mit diesen Donor-Funktionen ausgestattet. Mischkulturen von jenen Mutanten, die mit diesem Donor als Rezipient fungieren, können untereinander keine Rekombinanten hervorbringen. Abweichend von den durch HOPWOOD bei NF × UF-Kreuzungen des *Streptomyces coelicolor* gefundenen Verhältnissen wird die Donor-Eigenschaft des IMET JA 5142-01/6 nicht bzw. nicht obligatorisch auf die entstehenden Rekombinanten übertragen. Ebensowenig ist eine Hemmung des Rezipienten-Wachstums durch den Donor zu beobachten, wie sie der NF-Stamm beim UF-Stamm des *Streptomyces coelicolor* hervorruft.

Ob die Fertilitätseigenschaften bei den *Streptomyces griseus*-Mutanten von Plasmid-Genen determiniert sind, ist bisher nicht zu entscheiden. Allerdings konnte bisher weder bei den Wildstämmen noch bei ausgewählten Mutanten ein Plasmid nachgewiesen werden (GEUTHER, persönl. Mitteilung).

Bei allen fertilen Kreuzungen traten ausschließlich stabile Rekombinanten auf. Verschiedene Rekombinanten-Typen sind über 3 Generationen hinweg überprüft und dabei als stabil befunden worden. Heteroklone traten in den durchgeführten Kreuzungen nicht auf. Damit unterscheidet sich das hier vorgestellte Rekombinations-System in Anthracyclin-bildenden *Streptomyces-griseus*-Stämmen vom Rekombinations-System des Cephamycin-bildenden *Streptomyces griseus* NRRL 3851, in dem nach Kreuzungen ausschließlich Heteroklone gefunden wurden (PARAG 1978).

Bei der relativen Zuordnung der Marker wurde nach HOPWOOD (1969) eine zirkuläre Karte zugrunde gelegt, obwohl die vorliegenden Daten nicht ausreichen, diese Annahme zu bestätigen.

Die ersten Ergebnisse mit Mutanten Anthracyclin-bildender Streptomyceten-Stämme zeigen, daß unter speziellen Umweltbedingungen ein Rekombinations-System existiert, das für die genetische Analyse und Gen-Kartierung genutzt werden kann. Die bereits vorliegenden, Leukaemomycin-geblockten Mutanten des *Streptomyces griseus* (WAGNER *et al.* 1981) können für Untersuchungen zur genetischen Kontrolle der Anthracyclin-Biosynthese und deren gezielten Beeinflussung eingesetzt werden.

Wir danken Prof. MATSELYUKH, Kiew, für Hinweise und Diskussionen.

Literatur

ALACEVIC, M., 1969. Recombination in *Streptomyces* producing tetracyclins. In: Proceedings of International Symposium on Genetics and Breeding of *Streptomyces*, pp. 137–147. Zagreb: Yugoslav Academy of Sciences and Arts.
COATS, J. H. and ROESER, J., 1971. Genetic recombination in *Streptomyces bikiniensis var. zorbonensis*. J. Bacteriol., **105**, 880–885.
DELIC, V., HOPWOOD, D. A. and FRIEND, E. J., 1970. Mutagenesis by N-methyl-N-nitro-N-nitroso-guanidine in *Streptomyces coelicolor*. Mut. Res., **9**, 167–182.
FLECK, W. F. and STRAUSS, D., 1975. Leukaemomycin, an antibiotic with antitumor activity. I. Screening, fermentation and biological activity. Z. Allg. Mikrobiol., **15**, 495–503.
FRANCIS, M. M., CELLA, R. and VINING, L. C., 1975. Genetic recombination in a chloramphenicol-producing strain of *Streptomyces* species 3022a. Canad. J. Microbiol., **21**, 1151–1159.
FRIEND, E. J. and HOPWOOD, D. A., 1971. The linkage map of *Streptomyces rimosus*. J. gen. Microbiol., **68**, 187–197.
HOPWOOD, D. A., 1967. Genetic analysis and genome structure in *Streptomyces coelicolor*. Bacteriol. Rev., **31**, 373–403.
HOPWOOD, D. A., 1969. Genome topology and mapping in *Streptomyces coelicolor*. In: Proceedings of the International Symposium on Genetics and Breeding of *Streptomyces*. Dubrovnik, 1968. Yugoslav Academy of Sciences and Art, Zagreb, pp. 5–18.
HOPWOOD, D. A., HAROLD, R. J., VIVIAN, A. and FERGUSON, H. M., 1969. A new kind of fertility variant in *Streptomyces coelicolor*. Genetics, **62**, 461–477.
HOPWOOD, D. A., CHATER, K. F., DOWDING, I. E. and VIVIAN, A., 1973. Advances in *Streptomyces coelicolor* genetics. Bacteriol. Rev., **37**, 371–405.
HOPWOOD, D. A. and MERRICK, M. J., 1977. Genetics of antibiotic production. Bacteriol. Rev., **41**, 595–635.
KIRBY, R., 1978. Genetic mapping of *Streptomyces clavuligerus*. FEMS Microbiol. Lett., **3**, 177–180.
LEDERBERG, J. and LEDERBERG, E. M., 1952. Replica plating and indirect selection of bacterial mutants. J. Bacteriol., **63**, **399**.
MATSELYUKH, B. P., 1976. Structure and function of the *Actinomyces olivaceus* genome. In: K. D. MACDONALD (ed.), Second International Symposium on the Genetics of Industrial Microorganisms. Academic Press, London–New York and San Francisco. pp. 163–167.
PARAG, Y., 1978. Genetic recombination in *Streptomyces griseus*. J. Bacteriol., **133**, 1027–1031.
SERMONTI, G. and SPADA-SERMONTI, I., 1955. Genetic recombination in *Streptomyces*. Nature, **176**, 121.
WAGNER, CH., STENGEL, C., ERITT, I., SCHUMANN, G. und FLECK, W. F., 1981. Leukaemomycin-geblockte Mutanten des *Streptomyces griseus* und ihre Pigmente. Z. Allg. Mikrobiol., **21**, 751 bis 760.

Anschrift: G. SCHUMANN
Zentralinstitut für Mikrobiologie
und experimentelle Therapie der AdW
DDR 6900 Jena, Beutenbergstr. 11

Zeitschrift für Allgemeine Mikrobiologie	22	9	1982	649–660

(Department of Microbiology, St. Marianna University,
School of Medicine, Sugao, Miyamaye-ku, Kawasaki 213, Japan)

Effects of freezing and thawing and storage by freezing on the biological and morphological properties of strain K-9 of *Klebsiella pneumoniae* and its variants

M. Takahashi, S. Ito and K. Yoshida

(Eingegangen am 18. 2. 1982)

Using strain K-9 of *Klebsiella pneumoniae* and its variants A, B and C, which possess a large, small, or extra-small capsule or are unencapsulated, and exhibit high, middle, or low mouse virulence or avirulence in the mouse, respectively, the effect of freezing and thawing on the protection and storage by freezing in 20% skim milk solution on their biological and morphological properties was investigated. Experimental results showed that these organisms were sensitive against freezing and thawing in the order variant C, the parent strain, variants A and B. 20% skim milk solution was the most favourable for survival of the organisms compared to trypticase soy broth and 10% glycerol. Their mouse virulence was similarly sensitive to freezing and thawing experiments although no alteration of their biochemical properties was observed. When the organisms were suspended in 20% skim milk solution and stored for 2 years at −70 °C, their mouse virulence was not altered. When morphological features were tested electron microscopically, cellular injuries such as vacuola or fissions of cell walls were observed in the variants but not in the parent strain. However, it was evident that conversion to unencapsulation from encapsulation occurred at a high frequency in the parent strain and in variant A when determined by the colonial morphologies in soft-agar medium.

Although problems still remain in the storage of bacterial strains, investigations on the maintenance of biological properties of encapsulated strains are limited. With fresh isolates of *Klebsiella pneumoniae* labilities of biological characteristics involving high frequencies of variation have been noted (Edwards and Ewing 1972). Sakane and Sakano (1976) suggested difficulties of maintenance of these strains by drying due to wide range of viabilities (8 to 94%). Recently, we (Takahashi *et al.* 1977, 1978, Yoshida *et al.* 1981) described the biological and immunological properties of strain K-9 of *K. pneumoniae* and its variants A, B and C, which posses a large, small or extra-small capsule or an unencapsulated, respectively. Using these organisms the effects of freezing and thawing in the protection media and of storage by freezing on the biological and morphological properties were investigated and are described in this paper.

Materials and methods

Strain: Strain K-9 of *Klebsiella pneumoniae* and its variants A, B and C described elsewhere (Takahashi *et al.* 1977) were used. The parent strain and its variant A showing large round- and wedge-type colonial morphologies in soft-agar medium possessed large and small capsules and exhibited high and middle virulence in mice, respectively. Variants B and C possessed extra-small capsules or were unencapsulated, exhibited streaming- and compact-type growth in the medium, and had low virulence or avirulence in the mouse, respectively.

Procedure of freezing and thawing: The organisms cultured in trypticase soy broth (TSB) at 37 °C for 5 or 18 h were washed by centrifugation once with sterile phosphate buffered saline, pH 7.2 (PBS). The cells were resuspended in saline and the turbitities were adjusted nephelometrically to 2.0 OD at 430 nm. They were centrifuged, supernatants were discarded and the organisms were suspended in the original volume of TSB, 10% glycerol or 20% skim milk solution (DIFCO). These samples were rapidly frozen at −30 °C, then thawed at 30 °C. This treatment was repeated 5 times and the samples were obtained after every treatment.

Storage by freezing: 5 ml of the organisms suspended in 20% skim milk solution were frozen at −50 °C, then maintained at −70 °C deep freezer (Kelvinator) for 2 years.

Enumeration of viable cell number: The samples were suspended and diluted with PBS and viable cell numbers were enumerated by the regular plate count method.

Soft-agar technique: For the determination of colonial morphologies of the strains the soft-agar technique was performed as described (TAKAHASHI *et al.* 1977).

Determination of biochemical properties: The following products from EIKEN Co., Tokyo, were used to determine the biochemical properties of the parent strain and its variants: SIMMON's citrate agar, SIM medium, KLIGLER's agar, phosphate glucose peptone broth for VOGES-PROSKAUER and methyl red tests, malonate medium, urea broth, and KCN medium. DIFCO products included WORFEL-FERGUSON agar, DNase medium, ONPG broth, and assay media for lysine, arginine, and ornithine.

Determination of mouse virulence: The cell suspensions treated by freezing and thawing or stored in 20% skim milk solution at −70 °C for 2 years were serially diluted 10-fold with sterile saline, and 0.5 ml was injected intraperitoneally into a group of 5 mice. The number of dead animals was recorded for two weeks after the challenge injection.

Observation of morphological properties by the electron microscope: To the organisms suspended in 20% skim milk solution and maintained in a freezer 1.0 ml of 2.5% glutaraldehyde was added and kept at 4 °C for overnight. These samples were washed once with PBS, fixed again with 1% osmic acid for 2 hours at room temperature and embedded in Epon 812 according to the usual procedure after dehydration by ethanol and acetone. Ultra-thin sections were prepared with a PORTER-BLUM ultramicrotome, stained with uranyl acetate and lead citrate; then the organisms were observed by an electron microscope (Model 100B, Japan Optics Co. Ltd., Tokyo, Japan). Organisms which had not been frozen were used as controls.

Results

Survival of the strains after freezing and thawing

Primarily, the experiments were designed to elucidate whether the stability of the organisms against freezing and thawing differed depending upon the exponential or stationary growth phases.

Among the strains cultured for 5 hours and suspended in TSB, variant C showed the highest sensitivity, exhibiting no viable cells after 5 treatments. The viable cell number of the parent strain decreased by 3 orders of magnitude; a slight decline of the viable cell number was shown with the variants B and C. The survival of the organisms cultured for 18 hours did not change significantly, except for variant C where the viable cells decreased slightly, after repeated treatment as shown in Table 1.

Effects of protection media on the survival of the strains against freezing and thawing

When strains cultured for 18 hours at 37 °C were suspended in 20% skim milk solution and were repeatedly frozen and thawed, no significant decrease of viable cells was shown, except for variant C where viable cell counts slightly decreased. When the organisms were suspended in 10% glycerol the highest decrease was observed in variant C, and decreases of one order of magnitude were exhibited by variant B, however, no change appeared in the parent strain and variant A. In the TSB suspensions the lowest survival was also exhibited by variant C, and a slight decline was observed in the parent strain, although significant changes were observed in the variants A and B as shown in Table 2.

Table 1

Survival of strain K-9 *Klebsiella pneumoniae* and its variants (A, B and C) cultured 5 and 18 hours after freezing and thawing

Strain	Incubation period (hour)	Survival after freezing and thawing (time)					
		Before	1	2	3	4	5
K-9	5	1.9×10^9	6.2×10^8	1.1×10^8	3.9×10^7	8.0×10^6	1.2×10^6
	18	2.1×10^8	1.2×10^8	1.1×10^8	1.1×10^8	9.0×10^7	6.1×10^7
Variant A	5	7.8×10^8	3.0×10^8	8.0×10^7	6.1×10^7	5.6×10^7	3.5×10^7
	18	2.8×10^8	2.4×10^8	2.4×10^8	2.4×10^8	2.2×10^8	2.0×10^8
Variant B	5	6.2×10^8	5.5×10^8	3.6×10^8	8.9×10^7	6.2×10^7	4.5×10^7
	18	3.0×10^8	2.8×10^8	2.8×10^8	2.6×10^8	2.6×10^8	2.4×10^8
Variant C	5	1.3×10^9	1.2×10^5	7.4×10^2	3.0×10	0	0
	18	3.1×10^8	2.8×10^8	2.8×10^8	9.8×10^7	7.4×10^7	2.1×10^7

Table 2
Survival of strain K-9 of *Klebsiella pneumoniae* and its variants

Strain	suspended in	Survival after freezing and thawing (time)						stored for 2 years
		Before	1	2	3	4	5	
K-9	trypticase soy broth	2.1×10^8	1.2×10^8	1.1×10^8	1.1×10^8	9.0×10^7	6.1×10^7	ND
	10% glycerol	1.6×10^8	1.4×10^8	1.3×10^8	1.3×10^8	1.3×10^8	1.1×10^8	ND
	20% skim milk	2.2×10^8	2.0×10^8	2.0×10^8	2.0×10^8	1.8×10^8	1.2×10^8	5.6×10^7
Variant A	trypticase soy broth	2.8×10^8	2.4×10^8	2.4×10^8	2.4×10^8	2.2×10^8	2.0×10^8	ND
	10% glycerol	1.8×10^8	4.5×10^8	4.1×10^8	3.4×10^8	3.1×10^8	3.1×10^8	ND
	20% skim milk	2.1×10^8	2.1×10^8	2.0×10^8	2.0×10^8	2.0×10^8	1.9×10^8	9.1×10^7
Variant B	trypticase soy broth	3.0×10^8	2.8×10^8	2.8×10^8	2.6×10^8	2.6×10^8	2.4×10^8	ND
	10% glycerol	6.5×10^8	6.1×10^8	5.2×10^8	3.4×10^7	2.2×10^7	4.3×10^7	ND
	20% skim milk	2.5×10^8	2.5×10^8	2.4×10^8	2.4×10^8	2.5×10^8	2.5×10^8	7.8×10^7
Variant C	trypticase soy broth	3.1×10^8	2.8×10^8	2.8×10^8	9.8×10^7	7.4×10^7	2.1×10^7	ND
	10% glycerol	3.6×10^8	7.6×10^7	7.1×10^7	5.3×10^7	7.2×10^6	4.8×10^6	ND
	20% skim milk	3.0×10^8	2.4×10^8	2.2×10^8	2.2×10^8	1.8×10^8	1.2×10^8	8.1×10^6

Survival of the strains stored for two years by freezing

Since among the protection media 20% skim milk suspensions exhibited the best results for the survival of the organisms, exclusively this medium was tested for the effect of freezing for a long period of time on the organism. The organisms suspended in 20% skim milk solution were relatively stable stored for 2 years at —70 °C, except for the variant C whose survival slightly declined as shown in Table 2.

However, the colonial morphologies of the parent strain in soft-agar medium after 2 years showed a mixture of large round-, wedge-, streaming- to compact-type growth, although homologous colony size was observed when the strain was grown on MacConkey agar medium. The variant A showed three different colonial morphologies; large round-, wedge-, and streaming-type growth; however, the variants B and C did not change (Fig. 1).

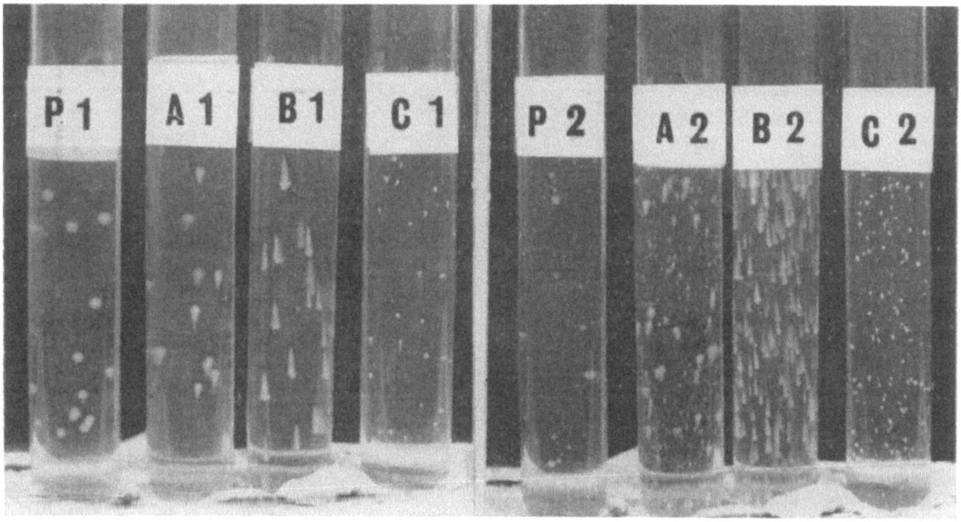

Fig. 1. Colonial morphologies of the K-9 strain of *Klebsiella pneumoniae* and its variants A, B and C in soft-agar medium.
P_1, A_1, B_1 and C_1 indicate the state before freezing of strain K-9 and its variants A, B and C, respectively. P_2, A_2, B_2 and C_2 are after 2 years storage of strain K-9, variants A, B and C in 20% skim milk solution at —70 °C, respectively

Biochemical properties of the parent strain and its variants

Strain K-9 and variants A, B and C showed similar biochemical reactions. They produced acid on the surface of Kligler's iron agar, except for variant C which showed an alkaline reaction on the surface; but all strains showed acid and gas production in the buttom, they all grew well on citrate, malonate, and KCN media, positive Voges-Proskauer reactions occurred, but not for indol, methyl red, and phenyl pyruvate. They were positive for galactosidase, urease, and the decarboxylase test with respect to lysine but not to arginine and ornithine; they were negative for DNase. All strains fermented sucrose and mannitol. After freezing and thawing and storage for 2 years in 20% skim milk solution, no alteration of these properties was observed in the parent strain as well as in the variants A, B and C.

Table 3

Sensitivities to freezing and thawing and frozen for 2 years on the mouse virulence of strain K-9 of *Klebsiella pneumoniae* and its variants

Strain	Suspended in	challanged with	Mouse virulence: Before storage dead/used	Mouse virulence: After storage dead/used
K-9	trypticase soy broth	10^3	5/5	0/5
		10^4	5/5	0/5
		10^5	5/5	0/5
	10% glycerol	10^2	5/5	0/5
		10^3	5/5	3/5
		10^4	5/5	5/5
	20% skim milk	10^2	5/5	3/5
		10^3	5/5	5/5
		10^4	5/5	5/5
	20% skim milk (after 2 years)	10^2	5/5	4/5
		10^3	5/5	5/5
		10^4	5/5	5/5
Variant A	trypticase soy broth	10^6	3/5	0/5
		10^7	5/5	0/5
		10^8	5/5	1/5
	10% glycerol	10^6	3/5	0/5
		10^7	5/5	3/5
		10^8	5/5	5/5
	20% skim milk	10^6	3/5	1/5
		10^7	5/5	5/5
		10^8	5/5	5/5
	20% skim milk (after 2 years)	10^6	3/5	3/5
		10^7	5/5	5/5
		10^8	5/5	5/5
Variant B	trypticase soy broth	10^6	0/5	0/5
		10^7	3/5	0/5
		10^8	5/5	0/5
	10% glycerol	10^6	0/5	0/5
		10^7	3/5	0/5
		10^8	5/5	0/5
	20% skim milk	10^6	0/5	0/5
		10^7	3/5	0/5
		10^8	5/5	2/5
	20% skim milk (after 2 years)	10^6	0/5	0/5
		10^7	3/5	0/5
		10^8	5/5	1/5
Variant C	trypticase soy broth	10^7	0/5	0/5
		10^8	0/5	0/5
	10% glycerol	10^7	0/5	0/5
		10^8	0/5	0/5
	20% skim milk	10^7	0/5	0/5
		10^8	0/5	0/5
	20% skim milk (after 2 years)	10^7	0/5	0/5
		10^8	0/5	0/5

Effect of freezing and thawing and storage by freezing on mouse virulence of the strains

To observe whether the mouse virulence of the organisms was altered by freezing and thawing and storage for long periods of time, the parent strain and its variants

A, B and C which showed high, middle, low, and no mouse virulence, respectively, were subjected to the experiments. The parent strain and the variants, both in the exponential and the stationary growth phase, suspended in distilled water, lost their mouse virulence after 5 cycles of freezing and thawing. When the organisms were suspended in 10% glycerol their virulence was also remarkably affected although the pathogenicities remained. In 20% skim milk solution, the parent strain was slightly affected by repeated freezing and thawing, but its virulence was still high. In this protection medium no significant alteration of the virulence of all strains occurred after storage for 2 years as shown in Table 3.

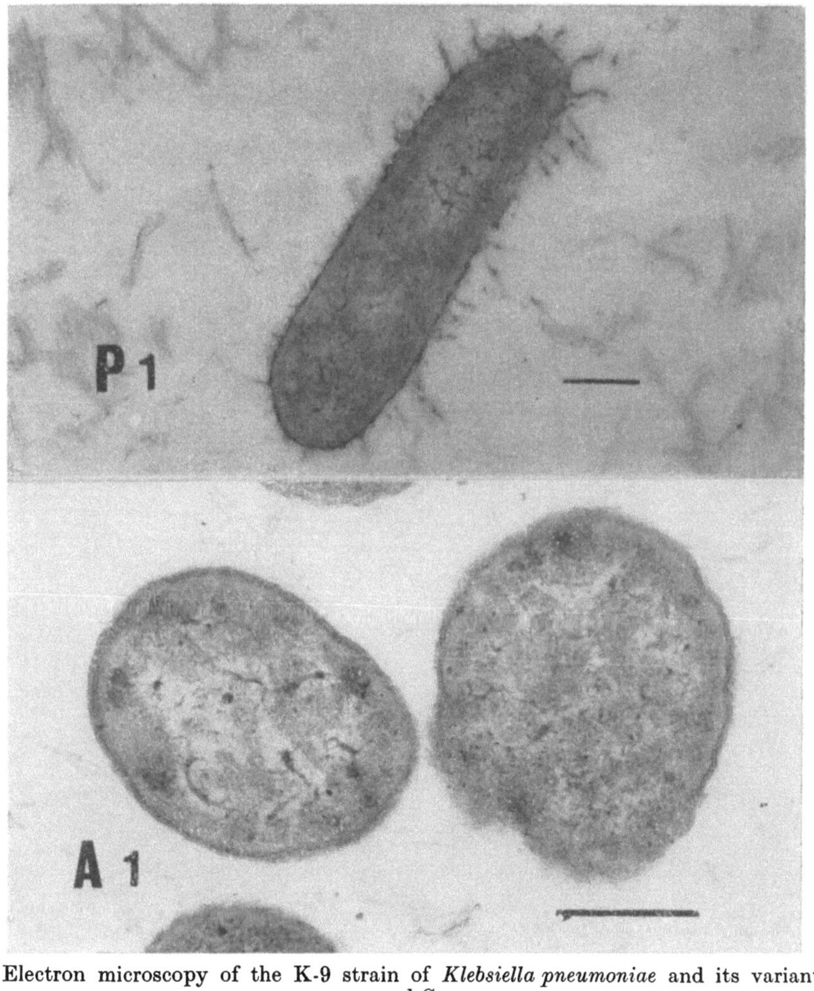

Fig. 2. Electron microscopy of the K-9 strain of *Klebsiella pneumoniae* and its variants A, B and C.
P_1, A_1, B_1 and C_1 are before freezing of strain K-9 and variants A, B and C, respectively. P_2, A_2, B_2 and C_2 are after 2 years storage of strain K-9 and variants A, B and C in 20% skim milk solution at -70 °C, respectively. The arrow indicates fissura of the cell wall

Electron microscopic observation of strains stored for 2 years

Since slightly decreasing survival was observed with strain K-9 and the variants suspended in 20% skim milk stored for long periods of time, they were observed electron microscopically. The parent strain suspended in 20% skim milk and stored for 2 years showed no significant morphological alterations compared to the control Fig. 2, A 2) exhibiting spike-like structures on the cell wall after the storage (Fig. 2, P 2). In variant A, numerous vacuols appeared within the cells and in variant B, fissura were observed in the cell wall. Variant C showed a number of vacuols (Fig. 2). These are presumed to be cellular injuries during storage in this protection medium.

Discussion

Detailed information concerning the effect of protection media for freezing and thawing on the biological and morphological properties of frozen organisms, and cell surface structures in particular, has been limited. In our series of papers, we emphasized a technique for determining the conversion from encapsulation to unencapsulation by the colonial morphologies in serum or plain soft-agar medium using the

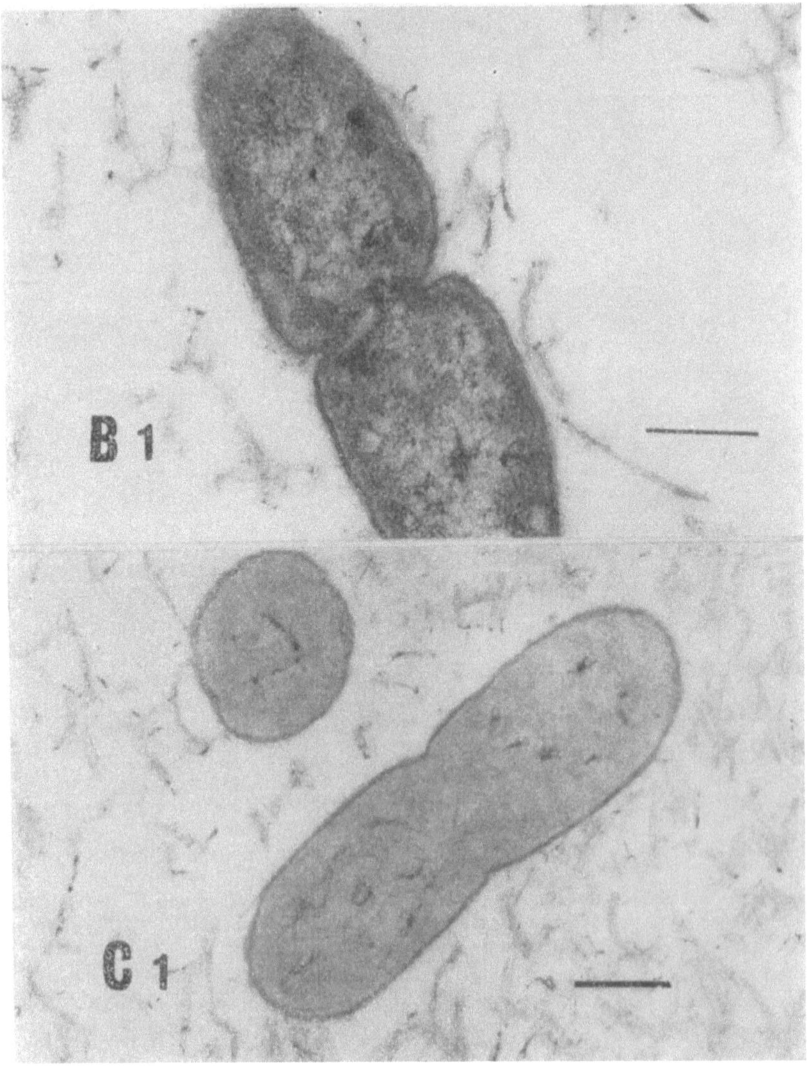

Fig. 2

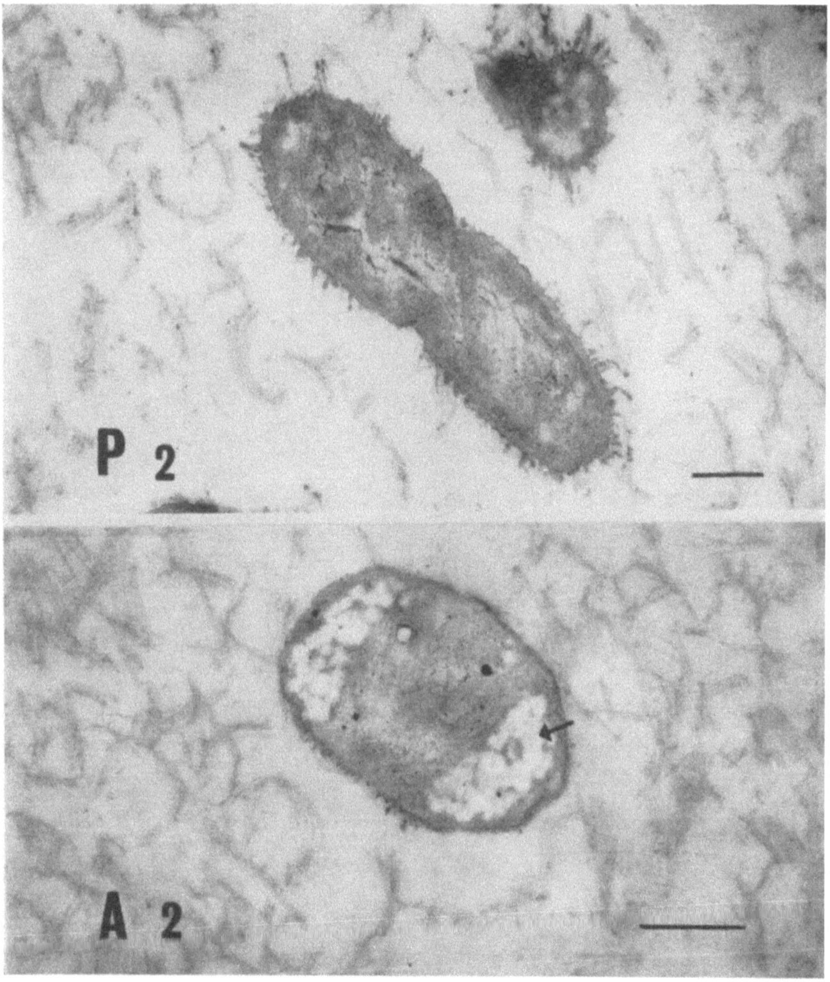

Fig. 2

following bacterial strains: *Staphylococcus aureus* (KOENIG 1962, YOSHIDA and TAKEUCHI 1970), *S. epidermidis*, group A *Streptococcus* (TAKAHASHI *et al.* 1980), *Str. pneumoniae* (TAKAHASHI *et al.* 1981), *Klebsiella ozaenea* (TAKAHASHI *et al.* 1978), and *Escherichia coli* (TAKAHASHI *et al.* 1980). Since distinct biological and morphological properties were demonstrated in strain K-9 of *K. pneumoniae* and its variants (TAKAHASHI *et al.* 1977, 1978) they were subjected to these experiments.

Concerning the effect of freezing and thawing on the survival of bacterial strains, higher sensitivities of organisms in the exponential phase than in the stationary phase have been observed in strains of *Pseudomonas aeruginosa* (GORRILL and MCNEIL 1960), *Serratia marcescens* (STRANGE and NESS 1963), and *Aerobacter aerogenes* (STRANGE and DARK 1962). In our experiments, the parent strain and its variants growing in the exponential phase were significantly more sensitive towards freezing and thawing compared to cells in the stationary phase. In particular, after 5 times of treatment no viable cells remained of variant C, an unencapsulated organism. The variants A and B which represent an intermediate type growth from encapsulated to unencapsulated showed the highest resistance against this treatment, followed by the

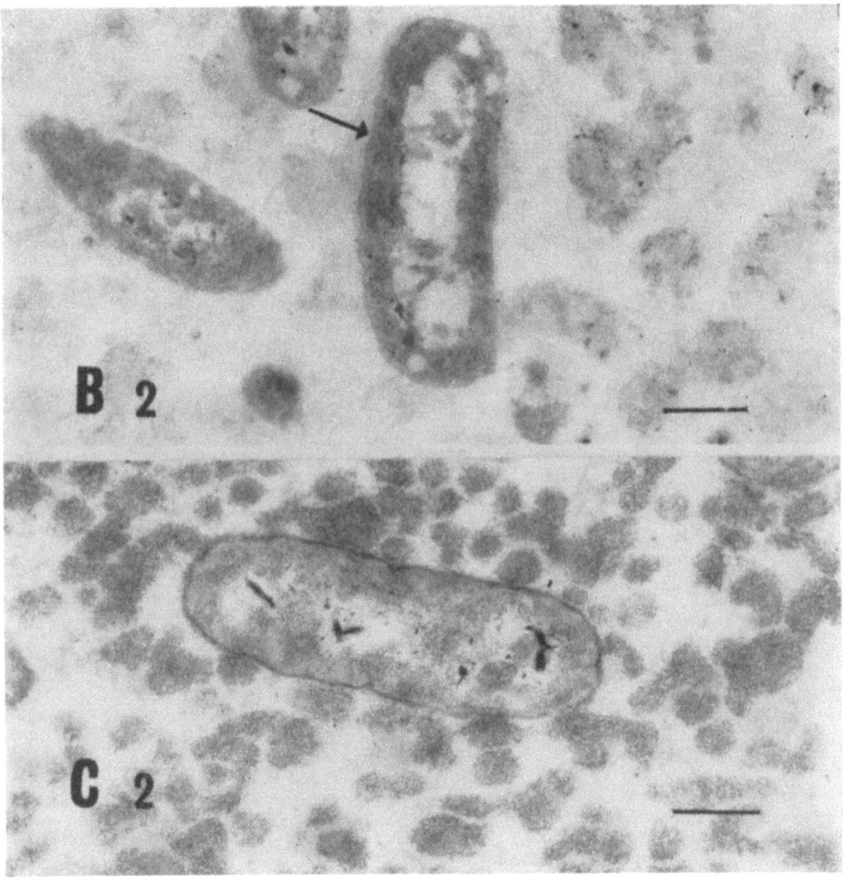

Fig. 2

parent strain although this organism possesses a large capsule. In the case of *S. aureus*, encapsulated organisms were significantly more sensitive against freezing than unencapsulated ones (OHTOMO *et al.* 1978). Our recent morphological findings partially confirmed these observations. However, the relation between different sensitivities to freezing and thawing and the state of the cell surface of these organisms is difficult to interpret. For protection against the lethal effect of the physical stimulation, dimethylsulfoxide (GIBSON *et al.* 1966), and glycerol (HOWARD 1956) have been used. Skim milk solution (WEISS 1957) has been widely applied in lyophilization, however, no paper concerned with its use for the storage by freezing is known to us. We found that the survival of variant C in the stationary phase decreased by one order of magnitude, however, no significant decrease of the viable cells was observed when they were suspended in 20% skim milk solution, indicating a satisfactory protective effect. During storage of the organisms suspended in 20% skim milk for 2 years a slight decrease of viable cells in the order of variant A, B, the parent strain and variant C occurred, similar to that in the freezing and thawing experiments. Regarding the biochemical properties of the parent strains and its variants, no alteration was observed after freezing and thawing and storage in 20% skim milk solution for 2 years.

Concerning mouse virulence, the strains suspended in TSB lost their virulence after freezing and thawing, however, organisms in 10% glycerol were slightly protected and mouse virulence of the organisms in 20% skim milk was very efficiently protected against freezing and thawing.

As to morphological changes of the strains stored for 2 years in 20% skim milk solution parent strain exhibited no alteration, however, a frequent change from encapsulation to unencapsulation could be interpreted by the soft-agar technique (TAKAHASHI *et al.* 1977), suggesting significant cellular injuries to have occurred in these organisms. In variant A, small size capsular organisms, numerous vacuols appeared within the cytoplasm, and fissura of the cell wall were observed in variant B. These are presumed to cellular injuries by freezing for long periods of time. The mechanism of sensitivity to freezing and thawing in relation to the structural differences of cell surfaces of the organisms, and the protective effect of skim milk solution on the biological properties of the bacterial strains against freezing and thawing require to be elucidated. These problems are currently under investigation in our laboratory.

References

EDMARDS, P. R. and EWING, W. H., 1972. Identification of *Enterobacteriaceae* (3rd Edition), p. 290—300. Burgess Publ. Co. Minneapolis/Minn.

GIBSON, C. A., LANDERKIN, G. B. and MORSE, P. M., 1966. Effects of additives on the survival of lactic streptococci in frozen storage. Appl. Microbiol., **14**, 665—669.

GORRILL, R. H. and MCNEIL, E. M., 1960. The effect of cold diluent on the viable count of *Pseudomonas pyocyanea*. J. gen. Microbiol., **23**, 437—442.

HOWARD, D. H., 1956. The preservation of bacteria by freezing in glycerol broth. J. Bacteriol., **71**, 625.

KOENIG, M. G., 1962. Factors relating to the virulence of Staphylococci. I. Comparative studies on the colonial variant. Yale J. Biol. Med., **34**, 537—559.

OHTOMO, T., YOSHIDA, K., IIZUKA, H. and SAN CLEMENTE, C. L., 1978. Relative stability of biological properties in freezing and encapsulated strains of *Staphylococcus aureus* after freezing-drying. Cryobiology, **15**, 461—468.

SAKANE, K. and SAKANO, I., 1976. Storage of bacterial strain by liquid-drying. Japanese Society for Research of Freeze and Freezing and Drying, **22**, 101—108.

STRANGE, R. E. and DARK, F. A., 1962. Effect if chilling on *Aerobacter aerogenes* in aqueous suspension. J. gen. Microbiol., **29**, 719—730.

STRANGE, R. E. and NESS, A. G., 1963. Effect of chilling on bacteria in aqueous suspension. Nature, **197**, 819.

TAKAHASHI, M., IWAMI, S. and YOSHIDA, K., 1978. Mouse virulence of a strain of *Klebsiella ozaenea* and its variants in relation to colonial morphologies in soft-agar. Microbiol. Immunol. **22**, 316—363.

TAKAHASHI, M., ICHIMAN, Y. and MARUOKA, Y., 1980. Relation of colonial morphologies of strains of group A streptococci in blood soft agar to morphological and biological properties. St. Marianna Med. J., **8**, 297—303.

TAKAHASHI, M., ICHIMAN, Y. and YOSHIDA, K., 1980. Selection of mouse virulent non-motile strains of *Escherichia coli* by the soft-agar technique. Acta microbiol. Acad. Sci. hung., **28**, 413—417.

TAKAHASHI, M., ICHIMAN, Y. and YOSHIDA, K., 1981. Relation of colonial morphologies of strains of *Streptococcus pneumoniae* in soft-agar to the encapsulation. St. Marianna Med. J., **9**, 229—234.

TAKAHASHI, M., YOSHIDA, K. and SAN CLEMENTE, C. L., 1977. Relation of colonial morphologies in soft-agar to morphological and biological properties of the K-9 strain of *Klebsiella pneumoniae* and its variants. Canad. J. Microbiol., **23**, 448—451.

TAKAHASHI, M., OIKAWA, M., KONO, E. and YOSHIDA, K., 1978. Electron-microscopic finding of surface structure and capsular swelling reaction of a strain of *Klebsiella pneumoniae*. Zbl. Bakteriol. Hyg., **240**, 497—502.

WEISS, F. A., 1957. Maintenance and preservation of cultures. In: Manual of Microbiological Methods (ed. by Society of American Bacteriologists), p 99–119. McGraw-Hill Publ. Co. New York.

YOSHIDA, K. and TAKEUCHI, Y., 1970. Comparison of compact and diffuse variants of the strains of *Staphylococcus aureus*. Infect. Immun., **2**, 523–527.

YOSHIDA, K., SMITH, M. R. and NAITO, Y., 1972. Serological typing of *Staphylococcus epidermidis* strains by the serum-soft agar technique. Infect. Immun., **5**, 8–11.

YOSHIDA, K., TAKAHASHI, M. and NARIKAWA, S., 1981. Amikacin in the treatment of mice with lethal infection. Experientia, **37**, 956–957.

Mailing address: Dr. K. YOSHIDA
Department of Microbiology,
St. Marianna University,
School of Medicine Sugao,
Miyamaye-ku, Kawasaki 213, Japan

Zeitschrift für Allgemeine Mikrobiologie	22	9	1982	661–670

(Akademie der Wissenschaften der DDR
Forschungszentrum für Molekularbiologie und Medizin,
Zentralinstitut für Mikrobiologie und experimentelle Therapie, Jena,
Direktor: Prof. Dr. U. TAUBENECK)

Elektronenoptische Untersuchungen zur *in vitro*-Wirkung von Dipyridamol auf das Pseudorabiesvirus

P. ZÖPEL

(Eingegangen am 4. 3. 1982)

Dipyridamole at a concentration of 50 µM/ml displays no activity on adsorption and penetration of pseudorabies virus in chicken embryonal cells. Furthermore, first stages of virus replication take place within the nucleus, whereas incomplete virus cores defective in DNA content are found within the nucleoplasm at times when the regular viral replication has been finished in controls. Defective pseudorabies virus particles lacking in DNA-content of the core, can be observed at the end of normal replication time. Consequently, the antiviral activity of dipyridamole may be due to blocking of the synthesis or of the incorporation of infectious viral DNA into the virus core.

Dipyridamol (DP) hemmt signifikant die Replikation einer Reihe von DNS- und RNS-Viren: Pox-, Herpeto-, Toga-, Orthomyxo-, Paramyxo- und Picornaviridae in unterschiedlichen Zellkulturen und im bebrüteten Hühnerei (TONEW *et al.* 1977). Da DP auch den transmembranösen Austausch von Nucleobasen inhibiert (SCHOLTISSEK 1968, PETERS u. HAUSEN 1971), könnte ein Einfluß auf die virale DNS-Synthese angenommen werden. Die vorliegende Arbeit untersucht die Wirkung von DP auf die Replikation des Pseudorabiesvirus (Herpetoviridae) in Hühnerembryonalfibroblasten, insbesondere bei frühen Replikationsschritten, entsprechend dem zeitlichen Ablauf der viralen DNS-Synthese.

Material und Methoden

Virus: Pseudorabiesvirus (PV), Stamm Aujeszky, Institut für Virologie der ČASV der ČSSR Bratislava; 40. Passage auf Hühnerembryonalfibroblasten (HEF).

Gewebekulturtechnik: HEF als Monolayerkultur; Aufarbeitung und Züchtung nach TONEW und TONEW (1969).

Dipyridamol (DP): 2,6-Bis(diäthanolamino)-4,8-dipiperidinopyrimidino-5,4-d-pyrimidin ist identisch mit der Aktivsubstanz des Präparates Curantyl 70 (GERMED DDR). Die Verdünnung des durch Umkristallisation gereinigten DP erfolgte mit absolutem Äthanol auf eine Stammkonzentration von 5 mM/ml, eine Arbeitskonzentration von 50 µM/ml wurde durch Zugabe von Gewebekulturmedium hergestellt.

Versuchsablauf: 3 Tage alte Kulturen von HEF in Petrischalen wurden nach 3maliger Spülung in einer phosphatgepufferten physiologischen Salzlösung (PBS) mit PV (10 PFU/Zelle) infiziert; Zugabe von 10 ml Gewebekulturmedium nach 30 min Adsorptionszeit bei 20 °C. Die DP-Konzentration blieb während des gesamten Versuchszeitraumes mit 50 µM/ml konstant. Substanzkontrollen und substanzfreie virusinfizierte Kontrollen wurden unter gleichen Bedingungen mitgeführt. Um den DP-Effekt auf den zeitlichen Ablauf der Virusreplikation fixieren zu können, wurde die Substanz während der Virusadsorption oder 30 min, 1, 2, 3, 4, 5, 7, 9, 12 bzw. 24 h nach der Infektion (p. i.) immer in gleicher Konzentration appliziert. Die Inkubation erfolgte bei 37 °C.

Elektronenmikroskopie: Fixierung der HEF *in situ* mit Glutaraldehyd-Osmiumsäure, Einbettung in Vestopal W (vgl. ZÖPEL *et al.* 1973). Herstellung der Ultradünnschnitte mit dem Ultramikrotom OM U2 (REICHERT, Wien); Kontrastierung nach REYNOLDS (1963). Aufnahme der Elektronenmikrogramme mit dem SIEMENS-Elmiskop I bei 80 kV.

Ergebnisse

Normaler Replikationszyklus

Übereinstimmend mit vorangegangenen Untersuchungen (ZÖPEL *et al.* 1973), beträgt die Replikationsdauer des PV in HEF *ca.* 12 h. Ein massiver cytopathogener Effekt (CPE) führt nach dieser Zeit zum Erlöschen der Zellfunktionen. Nach Virusadsorption- und -penetration treten 2—3 h post infectionem (p. i.) in einem überwiegenden Teil der Zellkerne, bevorzugt in elektronendichten Kernarealen, erste Replikationsstrukturen des Virus auf (Abb. 1). Sie haben einen Durchmesser von

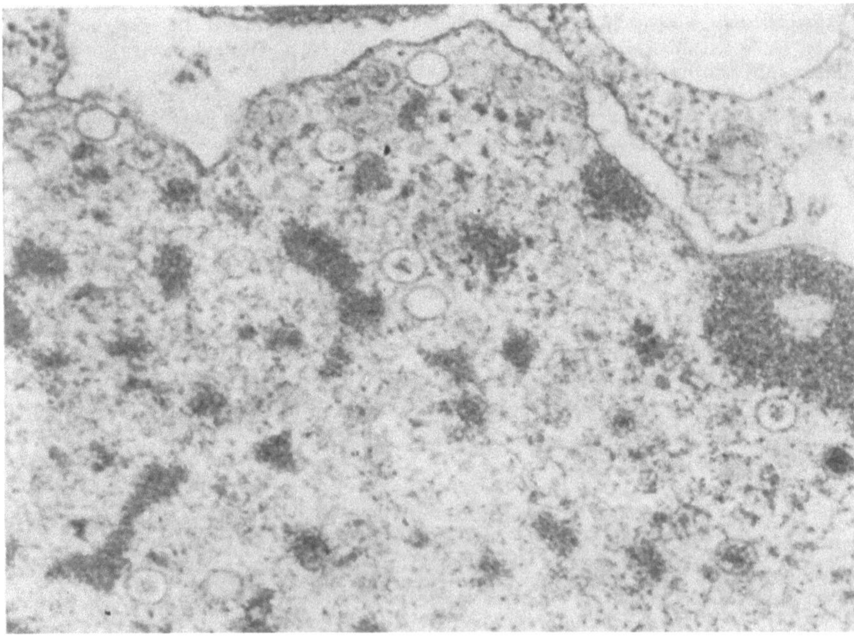

Abb. 1. Virus-cores als frühe Replikationsstrukturen in elektronendichten Arealen des Kerns einer HEF-Zelle, 3 h p. i. mit PV Vergr. 60000×

80—90 nm und sind isodiametrisch bis angedeutet hexagonal (Abb. 2). Ihre äußere Begrenzung ist elektronendicht und hat keinen Membrancharakter. Diese Partikeln erscheinen entweder leer oder lassen granuläre Binnenstrukturen unterschiedlicher Elektronenstreuung erkennen. 4 h p. i. besitzen $\geqq 80\%$ der Partikeln elektronendichte Kerne (Abb. 2). 6 h p. i. treten an der Innenseite der Kernmembran Aggregationen einzelner oder Gruppen dieser Partikeln auf, die von einem Membransystem umgeben sind. Dieser Vorgang leitet eine Ausschleusung aus dem Kern ein, die 7 h p. i. nahezu abgeschlossen ist. Gleichzeitig treten in der Nähe der Golgifelder reife Viruspartikeln in Vakuolen auf (Abb. 3). Mit ihrer Freisetzung aus der Zelle ist der Replikationszyklus beendet.

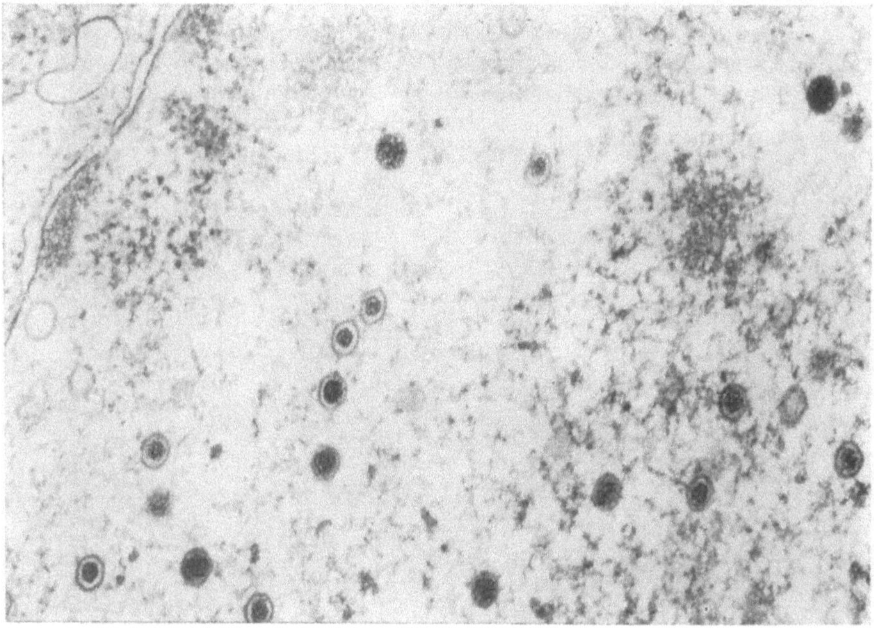

Abb. 2. Zahlreiche Virus-cores mit elektronendichtem Inhalt im Zellkern eines HEF, isodiametrische bzw. hexagonale core-Strukturen, 4 h p. i. Vergr. 45000×

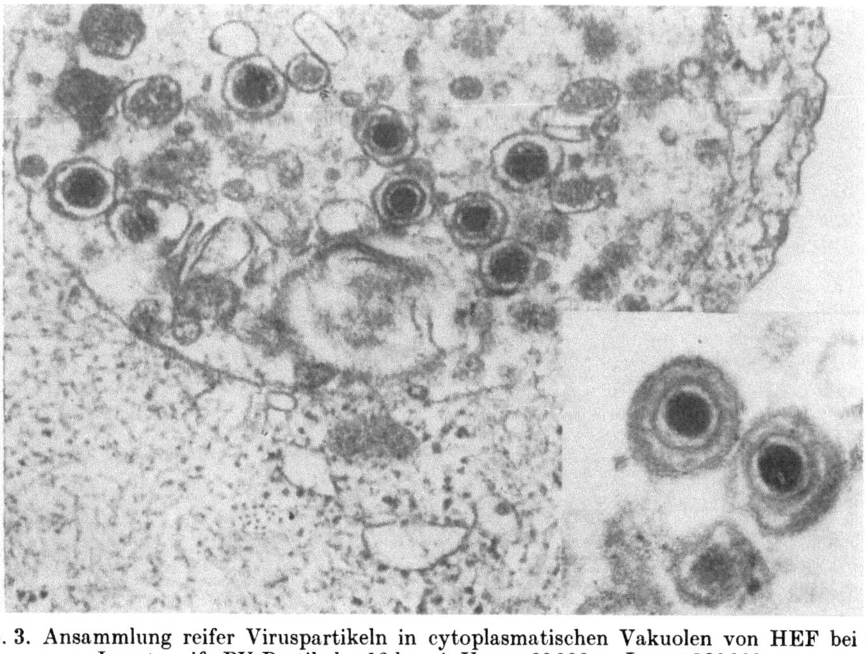

Abb. 3. Ansammlung reifer Viruspartikeln in cytoplasmatischen Vakuolen von HEF bei CPE. Insert: reife PV-Partikeln. 12 h p. i. Vergr. 60000×, Insert 120000×

Replikation unter Einwirkung von Dipyridamol

30 min p. i. werden Virusadsorptions- und -penetrationsstadien aufgefunden (Abb. 4, 5). Nach Durchlaufen der Eclipse (2—3 h p. i.) treten im Zellkern zahlreiche elektronendichte Areale als Frühstadien der Virusreplikation auf (Abb. 6). 4 h p. i.

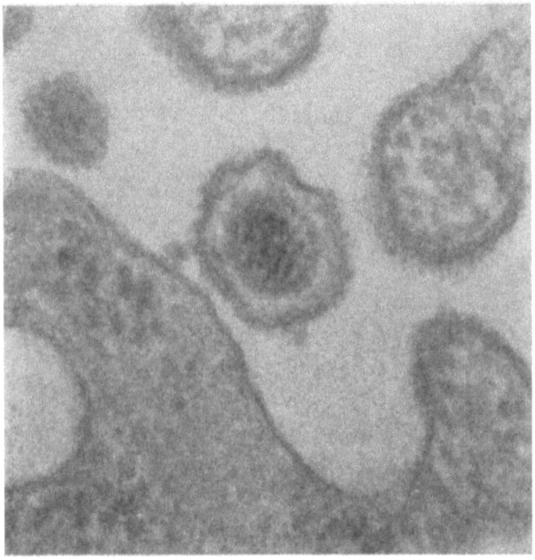

Abb. 4

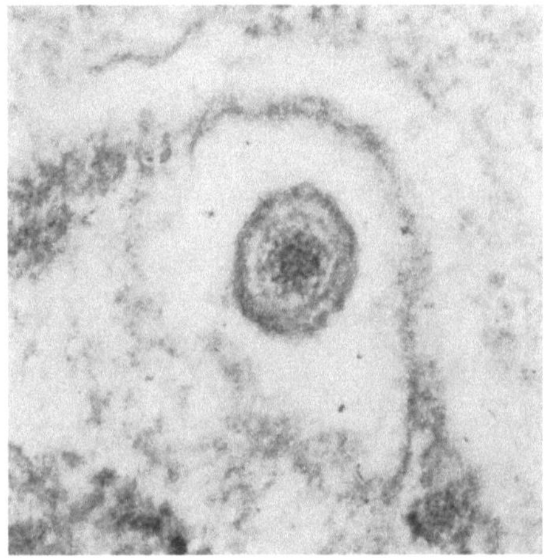

Abb. 5

Abb. 4. Adsorption von PV an der Zelloberfläche von HEF, 30 min p. i. und gleichzeitiger Inkubation mit 50 µM/ml DP. Vergr. 120000×

Abb. 5. Penetrationsstadium von PV in mikropinocytotischer Vakuole eines HEF, 30 min p. i. und gleichzeitiger Inkubation mit 50 µM/ml DP. Vergr. 120000×

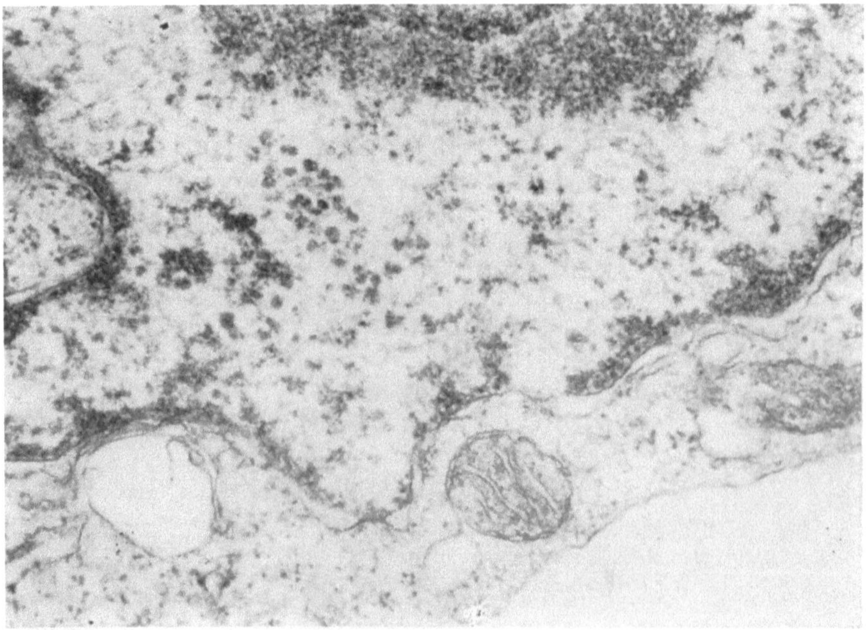

Abb. 6. Elektronendichte Strukturen als Frühstadien der Virusreplikation im Zellkern eines HEF, 4 h p. i. und gleichzeitiger Inkubation mit 50 µM/ml DP. Vergr. 45000×

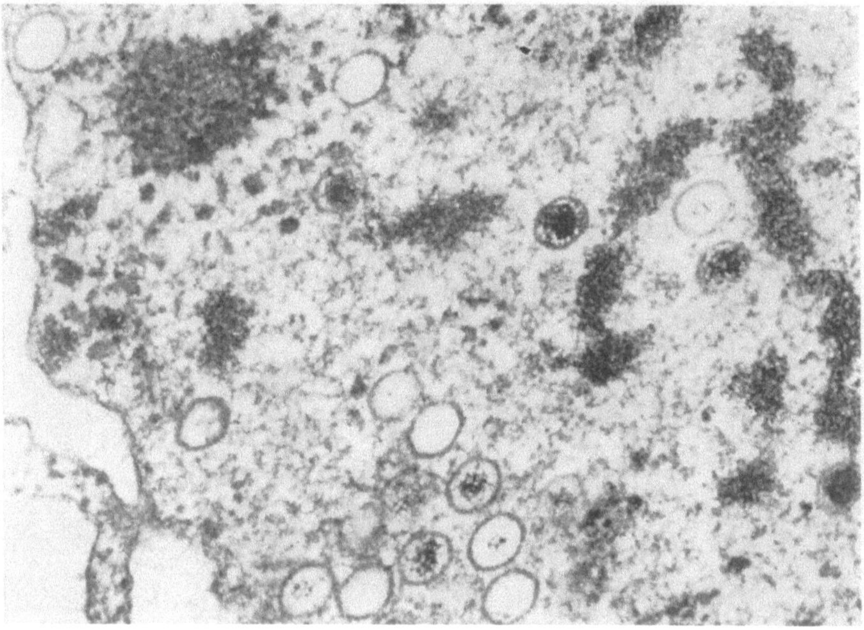

Abb. 7. Virus-cores mit überwiegend transparentem Zentrum im Zellkern eines HEF 4 h p. i. und gleichzeitiger Inkubation mit 50 µM/ml DP. Vergr. 80000×

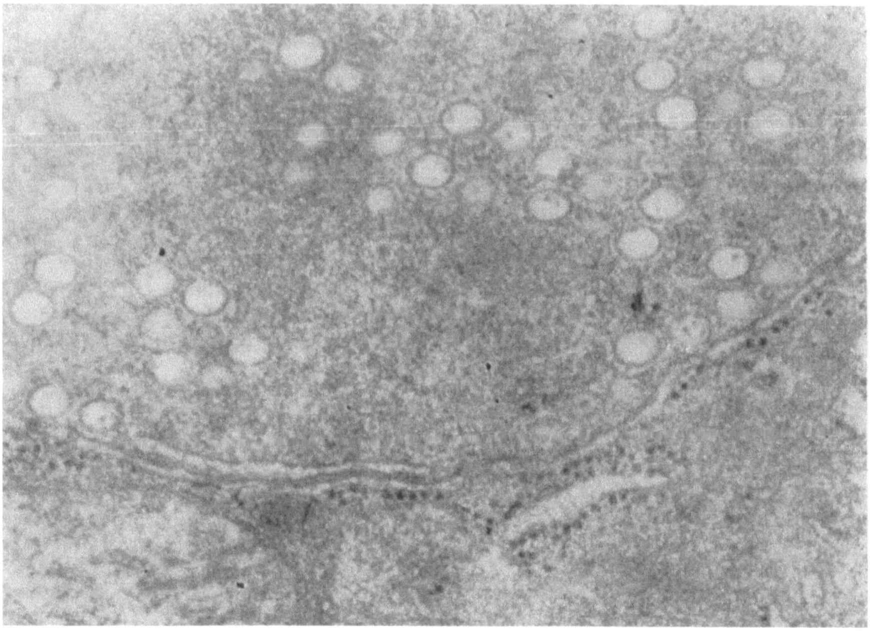

Abb. 8. Persistierende „leere" cores im Zellkern eines HEF, 12 h p. i. und gleichzeitiger Inkubation mit 50 µM/ml DP. Vergr. 60000×

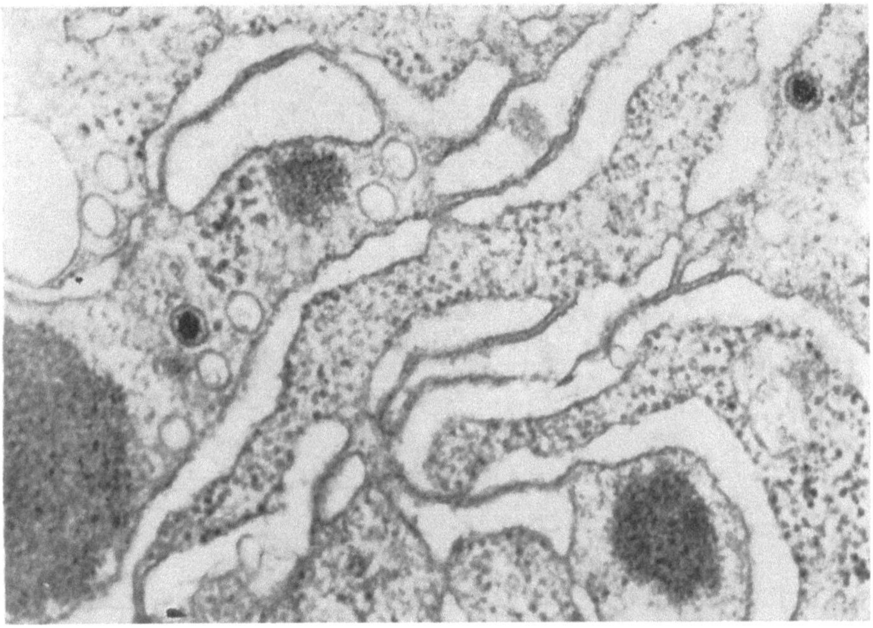

Abb. 9. Cytoplasmatische cores unterschiedlicher Elektronendichte in einem HEF, 9 h p. i. und gleichzeitiger Inkubation mit 50 µM/ml DP. Vergr. 60000×

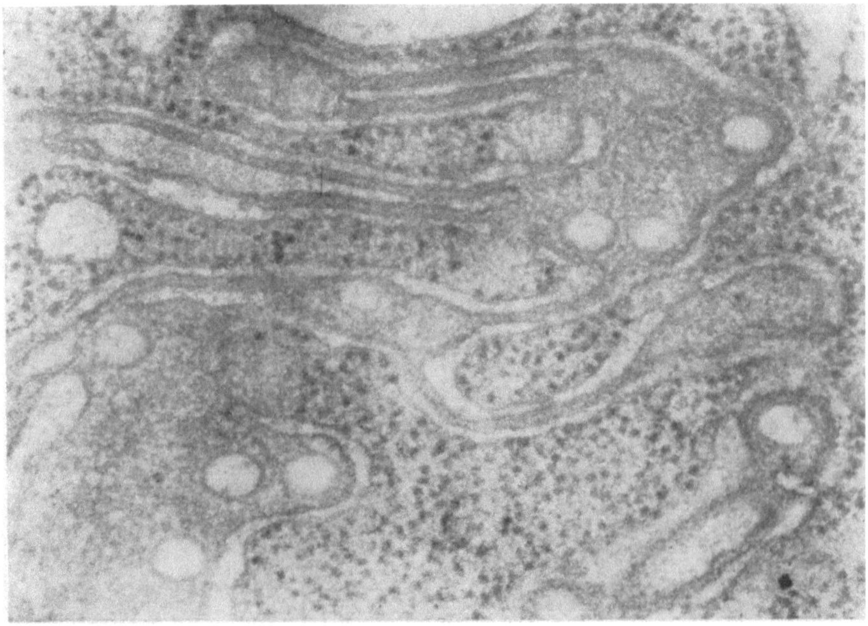

Abb. 10. Membransysteme bei Ausschleusung von PV-cores aus dem Kern eines HEF. Tangentialer Anschnitt eines HEF-Zellkerns mit Cytoplasma, 9 h p. i. und gleichzeitiger Inkubation mit 50 µM/ml DP. Vergr. 80000×

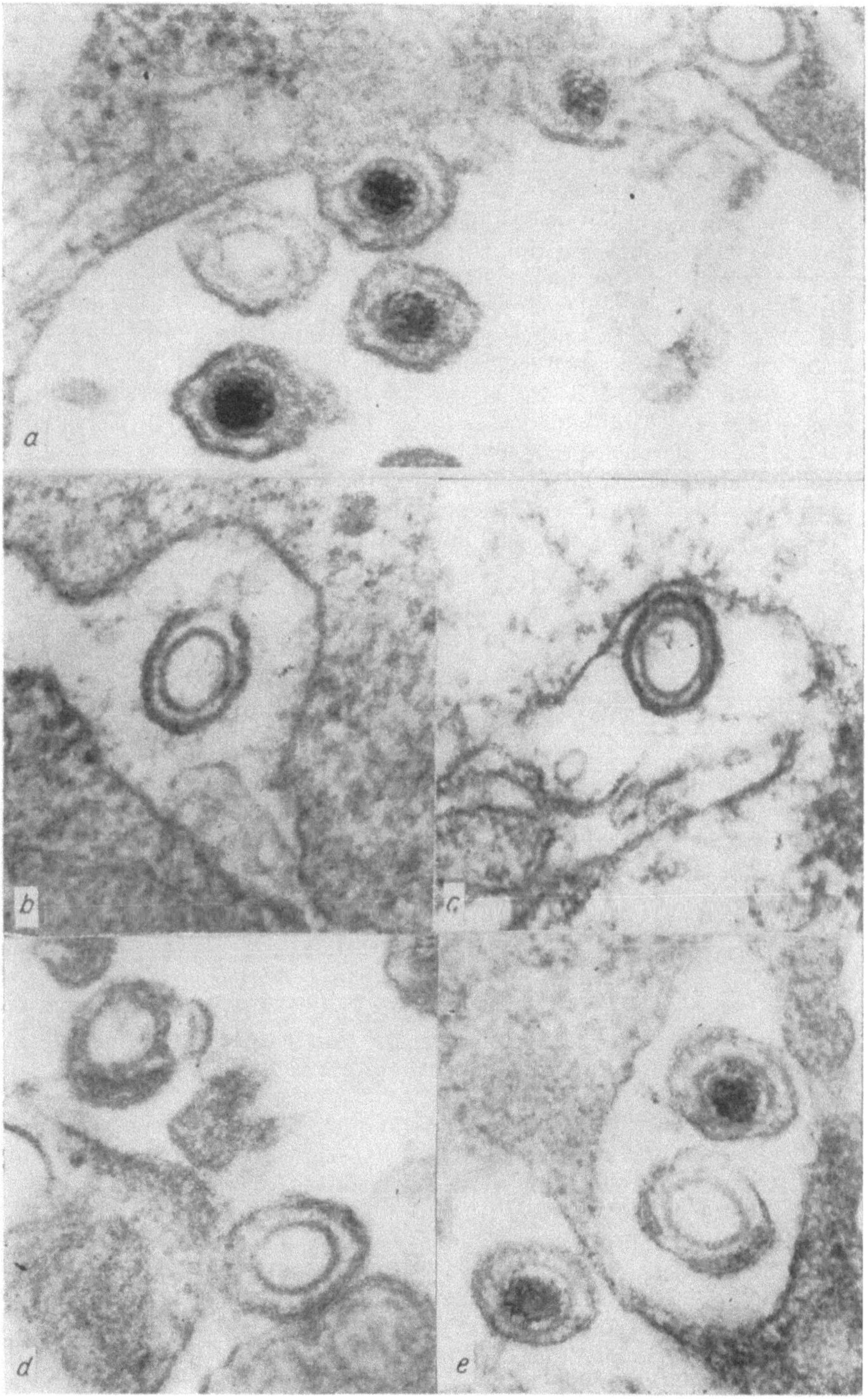

Abb. 11. Cytoplasma eines HEF, 12 h p. i. und gleichzeitiger Inkubation mit 50 μM/ml DP.
a: Umhüllung von PV-cores durch Knospung in cytoplasmatischer Vakuole
b—e: Reife PV-Partikeln mit z. T. „leeren" Virus-cores. Vergr. 120000×

lassen sich deutlich core-Strukturen erkennen (Abb. 7), die nach Abmessungen und morphologischem Status den Bildern der unbehandelten Kontrollen entsprechen, aber nach Auszählung eine geringere ($\leqq 55\%$) Anzahl aufweisen. Im Gegensatz zu Kontrollen gleichen Infektionsstadiums sind sie nicht oder nur partiell elektronendicht gefüllt, Auszählungen ergeben eine Reduktion um $\geqq 70\%$. Leere cores persistieren bis zu 12 h p. i. im Kern, wo sie teilweise dicht gepackt in umschriebenen Kernarealen auftreten (Abb. 8). 9 h p. i. können im Cytoplasma elektronenoptisch leere cores und solche mit unterschiedlich elektronendichtem Inhalt gefunden werden (Abb. 9). Ihre Ausschleusung (Abb. 10) aus dem Kern scheint nach dem gleichen Prinzip zu erfolgen, wie bei nichtbehandelten Zellen. Nach der 12. h p. i. können reife Viruspartikeln in cytoplasmatischen Vakuolen aufgefunden werden. Ihre cores besitzen unterschiedliche Elektronendichte, häufig treten „leere" cores auf (Abb. 11). Morphologische Zeichen eines CPE können unter Einfluß von DP bis 24 h p. i. nicht nachgewiesen werden. Eine Beeinflussung der PV-Vermehrung durch DP ist unter den gewählten Versuchsbedingungen nur bis 4 h p. i. möglich. Spätere Substanzgaben zeigen keine morphologische und chronologische Beeinflussung des replikativen Geschehens. Eine Titration der unter DP-Einwirkung gebildeten infektiösen Viruspartikeln im Virusertragstest ergibt eine Hemmung von 99,7%, eine Applikation von DP im gleichen Test 5 h p. i. oder später ergibt eine Hemmung von $\leqq 10\%$.

Diskussion

Das Auftreten von Replikationsstrukturen 2—3 h p. i. in HEF-Zellkernen schließt einen Einfluß von DP auf das Virus außerhalb der Zelle (Virucidie), seine Adsorption und Penetration aus (vgl. Oehring u. Schmidt 1978). Da eine DP-Applikation nach der 4. h p. i. die Virusreproduktion nicht eindeutig beeinflußt, muß der Zeitpunkt des Angriffs von DP zwischen Penetration und der 4. h des replikativen Geschehens liegen. Neben einer zeitlichen Verzögerung der Virusvermehrung und der Verkleinerung der Vermehrungsrate durch DP gegenüber Kontrollen ist ein quantitativer Anstieg von „leeren" cores von der 4. h p. i. an und deren Persistenz im Kern bis zur 12. h p. i. bemerkenswert. Der elektronendichte Zentralteil der Virus-cores stellt die replizierte PV-DNS dar. Pignatti und Cassai (1980) konstatieren bei Herpes simplex-Virus, einem nahen Verwandten des PV, eine direkte Korrelation zwischen DNS-Quantität und Elektronendichte des Virus-coreanteils. Cores mit elektronendurchlässigem Zentrum enthalten also keine oder nur Anteile von DNS. Aus ihnen resultierende, durch Virushüllproteine komplettierte Viruspartikeln, dürften demzufolge auch nicht infektiös sein.

Für das Entstehen DNS-freier Virus-cores können u. a. Mutationen verantwortlich sein. Atkinson *et al.* (1978) beschreiben eine thermosensitive „empty"-Mutante bei Herpes simplex-Virus. Bei hohen Vermehrungsraten nach hochtitriger Beimpfung des Zellrasens (mehr als 100 PFU/cell) treten häufig DNS-Defekte bei Herpetoviridae auf. Ben-Porat *et al.* (1974) und Rubenstein und Kaplan (1975) konnten nach unverdünntem Passagieren von PV eine Anzahl von Viruspartikeln mit DNS-Defekten nachweisen. Das gleiche Phänomen wurde von Bronson *et al.* (1973) und Becker *et al.* (1978) bei Herpes simplex-Virus beobachtet und von letzteren mit einer Fragmentation der linearen Virus-DNS in zirkuläre Teilstücke erklärt, wobei für die DNS-Reproduktion ein „rolling circle mechanism" postuliert wird. Von den meisten vorgenannten Autoren wird für die DNS-Defekte eine nicht ausreichende Syntheseleistung der Wirtszelle für virusspezifische DNS verantwortlich gemacht; häufig werden nur DNS-Fragmente in die synthetisierten Viruscores eingebaut. Wenn unter

DP-Einfluß zahlreiche DNS-freie PV-cores auftreten, so kann entweder der Einbau der viralen DNS in die cores gestört sein oder ein Defekt in der virusspezifischen DNS-Synthese vorliegen.

Durch DP wird nach SCHOLTISSEK (1968) die Aufnahme der DNS-Bausteine Adenosin, Cytidin und Guanosin in Gewebekulturzellen sowie die Permeation von Thymidin und Desoxyadenosin in die Zelle gehemmt. KESSEL und HALL (1970) beschreiben einen Hemmeffekt des DP auf den Desoxycytidintransport durch zelluläre Membransysteme.

TONEW und DZEGUZE (1977) sowie TONEW *et al.* (1978) beschreiben nach DP-Gabe eine Replikationshemmung des Mengo-Virus in FL-Zellen, die auf eine Hemmung der virusspezifischen RNS-Synthese bezogen wird. Ihre Ursache ist ein verminderter Uridintransport in die Zelle. Der gleiche DP-Effekt wird auch von PLAGEMANN und ROTH (1969) sowie PETERS und HAUSEN (1971) bei der Untersuchung anderer Zellsysteme beschrieben.

Ein durch DP induzierter Mangel an Nucleinsäureprecursoren im Zellinnern würde an den Nucleinsäuresyntheseorten zu einem verringerten Angebot von Nucleosiden und damit zu einer geringeren DNS-Syntheserate führen.

Die Tatsache, daß in DP-behandelten Zellen auch eine Anzahl morphologisch intakter PV-Elementarkörper erscheint, der Virusertragstest aber eine Hemmung der Infektiosität von 99,7% anzeigt, läßt den Schluß zu, daß ein großer Teil auch dieser Viren DNS-Defekte aufweist. Die vorliegenden elektronenoptischen Befunde unterstützen die Annahme eines Defektes bei der viralen DNS-Synthese oder bei deren Einbau in die vorliegenden Virus-cores, wobei für ersteres die Beeinflussung der Replikation durch DP-Gabe nur bis 4 h p. i. spricht.

Den Herren Prof. G. BRUNS und Dr. med. habil. S. FRITSCH danke ich für die kritische Durchsicht des Manuskripts und Frau IRENE LEPPER für die sorgfältige Ausführung der Präparationsarbeiten.

Literatur

ATKINSON, M. A., BARR, S. and TIMBURY, M. C., 1978. The fine structure of cells infected with temperature sensitive mutants of herpes simplex virus type 2. J. gen. Virol., **40**, 103–119.

BECKER, J., ASHER, Y., WEINBERG-ZAHLERING, E., RABKIN, S., FRIEDMAN, A. and KESSLER, E., 1978. Defective herpes simplex virus DNA: Circular and circular-linear molecules resembling rolling circles. J. gen. Virol., **40**, 319–335.

BEN-PORAT, T., DEMARCHI, J. M. and KAPLAN, A. S., 1974. Characterization of defective interfering viral particles present in a population of pseudorabies virus. Virology, **61**, 29–37.

BRONSON, D. L., DREESMAN, G. R., BISWAL, N. and BENYESH-MELNICK, M., 1973. Defective virions of herpes simplex virus. Intervirology, **1**, 141–153.

KESSEL, D. and HALL, T. C., 1970. Effects of persantin on deoxycytidine transport by murine leukemia cells. Biochim. biophysica Acta, **211**, 88–94.

OEHRING, H. und SCHMIDT, J., 1978. Zur antiviralen Wirkung von Dipyridamol auf Rhinoviren. Z. ges. Hyg., **24**, 447–449.

PIGNATTI, P. F. and CASSAI, E., 1980. Analysis of herpes simplex virus nucleoprotein complexes extracted from infected cells. J. Virol., **36**, 816–828.

PLAGEMANN, P. G. W. and ROTH, M. F., 1969. Permeation as the rate limiting step in phosphorylation of uridine and choline and their incorporation into macromolecules by NOVIKOFF hepatoma cells. Competitive inhibition by phenetyl alcohol, persantin, and adenosine. Biochemistry, 8, 4782–4789.

REYNOLDS, E. S., 1963. The use of lead citrate at high pH as an electron opaque stain in electron microscopy. J. Cell Biol., **17**, 208–212.

RUBENSTEIN, A. S. and KAPLAN, A. S., 1975. Electron microscopic studies of the DNA of defective and standard pseudorabies virions. Virology, **66**, 385–392.

Scholtissek, C., 1968. Studies on the uptake of nucleic acid precursors into cells in tissue culture. Biochim. biophysica Acta, **158**, 435–447.

Tonew, E. und Tonew, M., 1969. Die Wirkung von Cycloheximid auf das Virus Abortus der Schafe und Sindbis Virus, sowie andere Viren in Screening-Untersuchungen. Zbl. Bakteriol., I. Abt., Orig., **211**, 437–444.

Tonew, M. and Dzeguze, D., 1977 Dipyridamole, an inhibitor of Mengovirus replication in FL and L cells. Chemotherapy, **23**, 149–158.

Tonew, M., Tonew, E. and Mentel, W., 1977. The antiviral activity of dipyridamole. Acta virol., **21**, 146–150.

Tonew, M., Laass, W., Tonew, E., Franke, R., Goldner, H. and Zschiesche, W., 1978. Antiviral activity of dipyridamole derivatives. Acta virol., **22**, 287–295.

Zöpel, P., Eckardt, K. und Tonew, E. M., 1973. Elektronenoptische Untersuchungen zur Wirkung des Streptomycetenantibioticums Borrelidin auf das Pseudorabiesvirus. Z. Allg. Mikrobiol., **13**, 711–722.

Anschrift: Dr. P. Zöpel
Zentralinstitut für Mikrobiologie und
experimentelle Therapie der AdW
DDR 6900 Jena, Beutenbergstr. 11

Zeitschrift für Allgemeine Mikrobiologie	22	9	1982	671–673

Kurze Originalmitteilung

(Wissenschaftsbereich Mikrobielle Biochemie der Sektion Biologie
der Friedrich-Schiller-Universität, Jena)

In vitro-Einbau von ^{14}C-Hexose-6-phosphat in Mannan, β-Glucan und Glycogen bei *Candida spec.* H und ihren Mutanten

B. RÖBER und G. REUTER

(Eingegangen am 22. 4. 1982)

Mannose-6-p is an activator of the ^{14}C-mannose incorporation from GDP-^{14}C-mannose in the mono- and oligosaccharides and in the mannopolymers of the cell wall proteophosphomannan produced by the food protein yeast *Candida spec.* H.
Moreover, mannose-6-p is a precursor of proteophosphomannan: ^{14}C-Mannose-6-p has been incorporated in absence of GTP. Corresponding behaviour shows glucose-6-p by synthesis of β-glucan and glycogen. Mutants of *Candida spec.* H with different efficiency in the biosynthesis of mannan, β-glucan and glycogen incorporate hexose-6-p in a different extent.

In Verbindung mit Untersuchungen über die Struktur und Biosynthese von Zellwand- und Reserve-Polysacchariden an *Candida spec.* H, einer Kohlenwasserstoff-verwertenden Hefe (RÖBER u. REUTER 1979, GRIMMECKE u. REUTER 1980, 1981), wurden Untersuchungen zur Regulation und Precursor-Bereitstellung aufgenommen, die bisher fast ausschließlich an *Saccharomyces cerevisiae* durchgeführt wurden (z. B.: BEHRENS u. CABIB 1968, NAKAJIMA u. BALLOU 1974 und 1975, BECKER 1977, LEHLE u. TANNER 1978, Übersicht bei BALLOU 1976 und FARKAS 1979).

Aus *Candida spec.* H wurden Mutanten mit veränderter Polysaccharid-Biosynthese gewonnen und auf Glucose-Harnstoff-Minimalmedium aerob kultiviert. Aus Zellen der logarithmischen Wachstumsphase wurden die spezifischen Polysaccharid-Syntheseleistungen ermittelt: Glucan-Synthase und Glycogen-Synthase in zellfreien Homogenaten. Mannan-Synthase im 105000 × g Sediment, radiometrischer Test in Anlehnung an BEHRENS und CABIB (1968). Neben den Substraten des Glycosyltransfers (Tab. 1) wurden die in den Enzymansätzen vorhandenen nativen Polysaccharide als Akzeptormoleküle verwendet. Die erhaltenen radioaktiven Polysaccharide wurden enzymatisch (zur Trennung von Glycogen- und β-Glucan) oder alkalisch (zur Trennung von Mono- und Oligosacchariden von Proteophosphomannan durch Spaltung der Esterglycoside und Ser/Thr/O/-Glycoside) abgebaut und der ^{14}C-Einbau aus den markierten Precursoren in die erhaltenen Abbau-Produkte bestimmt.

Nach Tabelle 1 aktiviert bei *Candida spec.* H Mannose-6-Phosphat den Einbau von ^{14}C-Mannose aus GDP-^{14}C-Mannose in die Mono- und Oligosaccharide (Faktor 22) und in die Manno-Polymere (Faktor 2,8). Bei Einsatz von ^{14}C-Mannose-6-P neben nichtradioaktiver GDP-Mannose wird deutlich, daß ^{14}C-Mannose-6-P nicht nur Aktivator, sondern auch Precursor der Mannan-Biosynthese ist. So werden bei der synthetisierten alkalilabilen Mono- und Oligosaccharid-Fraktion des Mannans 41% der incorporierten Hexose ($0{,}82 + 0{,}58 = 1{,}4$ nmol · min^{-1} · mg^{-1} = 100%) aus Mannose-6-P bereitgestellt, beim synthetisierten Manno-Polymer sogar 91% ($0{,}19 + 1{,}92 = 2{,}1$ nmol · min^{-1} · mg^{-1} = 100%).

In weiterführenden Untersuchungen an *Candida spec.* H konnten wir feststellen, daß in Abhängigkeit von Mannose 6-P bevorzugt der Einbau der ersten Seitenketten-

Tabelle 1

Polysaccharid-Biosyntheseleistung *in vitro* von *Candida spec.* H und deren Mutanten H3 und H5 (nmol · min^{-1} · mg^{-1} Protein; Mittelwerte aus zwei Präparationen mit je vier Parallelproben)

Enzym	Substrat	spezifische Enzym-Aktivität		
		H	H3	H5
Mannan-Synthase (Mono- und Oligosaccharide)	GDP-^{14}C-Man	0,038	0,038	0,042
	GDP-^{14}C-Man + UDP-Glc	0,038	0,031	0,033
	GDP-^{14}C-Man + Man-6-P	0,82	0,53	0,75
	GDP-Man + ^{14}C-Man-6-P	0,58	0,29	0,73
Mannan-Synthase (Manno-Polymer, MWG 1.1 · 10^5)	GDP-^{14}C-Man	0,068	0,085	0.139
	GDP-^{14}C-Man + UDP-Glc	0,068	0,051	0.095
	GDP-^{14}C-Man + Man-6-P	0,19	0.14	0.28
	GDP-Man + ^{14}C-Man-6-P	1‘92	1,01	2.10
Glycogen-Synthase	UDP-^{14}C-Glc	0,69	1,20	0,64
	UDP-^{14}C-Glc + Glc-6-P	5,2	7,5	1,97
	UDP-Glc + ^{14}C-Glc-6-P	20,1	12,4	2,56
Glucan-Synthase	UDP-^{14}C-Glc	0,050	0,058	0,046
	UDP-^{14}C-Glc + Glc-6-P	0,51	0,58	0,23
	UDP-Glc + ^{14}C-Glc-6-P	8,9	6,5	15,3

Mannose erfolgt unter Ausbildung einer α(1,2)-resp.α(1,3)-Bindung (vgl. Strukturmodell Proteophosphomannan, Grimmecke u. Reuter 1980).

Durch Abbau des *in vitro* aus 1-^{3}H-Mannose-6-P synthetisierten radioaktiven Mannans konnten wir feststellen, daß ein alkalilabil gebundenes Oligosaccharid mit dem Polymerisationsgrad 6—7 (MG 1080—1260) aus Mannose-6-P entsteht, während die alkalilabil gebundenen Mannose- und Dimannose-Einheiten bevorzugt aus der gleichzeitig angebotenen GDP-U-^{14}C-Mannose gebildet werden (Röber u. Reuter 1980).

Wie Tabelle 1 zeigt, ist die *in vitro* Synthese von Glycogen sowie Glucan durch Glucose-6-Phosphat in ähnlicher Weise aktivierbar, wie dies für andere Hefearten bereits nachgewiesen wurde (Becker 1977). Darüber hinaus sprechen jedoch unsere Ergebnisse dafür, daß ^{14}C-Glucose-6-P entsprechend dem ^{14}C-Mannose-6-P ein Precursor der Polysaccharide sein kann: 79% des neu synthetisierten Glycogens und 95% des neu synthetisierten Glucans werden von Glucose-6-P geliefert.

Ein Mechanismus für die Polysaccharid-Biosynthese aus Hexose-6-Phosphaten unter den vorliegenden Bedingungen wurde bisher nicht beschrieben. Der Weg über Hexose-1-P und UDP-Glc resp. GDP-Man ist bei Abwesenheit von UTP resp. GTP nicht plausibel, es sei denn, Glucose aus UDP-Glc wird in einer Austauschreaktion gegen ^{14}C-Glc-6-P und Mannose von GDP-Man gegen ^{14}C-Man-6-P ausgewechselt. Andernfalls müßten Einschleusungsreaktionen unter Umgehung einer Aktivierung durch UTP resp. GTP postuliert werden. Ferner wäre in die Betrachtung die Aktivierung der Hexose-Phosphate über eine Dolichyl-Phosphat/Dolichyl-Hexose-Phosphat-Sequenz (Lehle *et al.* 1980) einzubeziehen. Durch das Postulat einer Austauschreaktion wäre ein Anschluß möglich an die Arbeiten über die Beteiligung einer Mannosyl-P-Transferase an der Synthese von Phosphomannan (Thieme *et al.* 1971, Slodki 1962, 1963, Mayer 1971).

Wie Tabelle 1 zeigt, sind auch die Mutanten von *Candida spec.* H in der Lage, Hexose-6-P unter Ausschluß von UTP resp. GTP in die entsprechenden Polysaccharide zu incorporieren. Allerdings treten zwischen den Stämmen erhebliche Unterschiede auf. Dies spiegelt sich wider in den unterschiedlichen Mannan-, β-Glucan- und Glycogen-Biosyntheseleistungen bei den von uns *in vivo* durchgeführten Versuchen.

Aus den Versuchsergebnissen kann geschlossen werden, daß sich Hefe-Mutanten in besonderer Weise zur Biomasseproduktion auf Kohlenhydraten eignen, die sich durch eingeschränkte Möglichkeit zur Biosynthese von Polysacchariden aus Hexose-6-P auf den von uns postulierten Wegen auszeichnen und die zusätzlich eine Repressionsdesensibilisierung des aeroben Kohlenhydratstoffwechsels aufweisen, da aerob bei hohen Wachstumsraten und hohem Y_c eine günstige Biomassezusammensetzung realisiert werden kann.

Auch für Kohlenwasserstoff assimilierende Hefen dürften nach den Ergebnissen von Käppeli und Fiechter (1977) zum Prozeß der Aufnahme von hydrophoben C- und Energiequellen Hefe-Mutanten von Bedeutung sein, die bezüglich Anteil und Struktur der Zellwand-Polysaccharide Veränderungen zeigen.

Literatur

Ballou, C. E., 1976. Structure and biosynthesis of the mannan component of the yeast cel envelope. Adv. Microbiol. physiol., **14**, 93–158.

Becker, J. U., 1977. Glycogen synthase in *Saccharomyces carlsbergensis* cells capable of oscillatory glycolysis. Arch. Microbiol., **115**, 181–184.

Behrens, N. H. and Cabib, E., 1968. The biosynthesis of mannan in *Saccharomyces carlsbergensis.* J. biol. Chemistry, **243**, 502–509.

Farkas, V., 1979. Biosynthesis of cell walls of fungi. Microbiol. Rev., **43**, 117–144.

Grimmecke, H.-D. und Reuter, G., 1980. Struktur der Zellwandpolysaccharide in der Futtereiweißhefe *Candida spec.* H. V. Die komplexe Struktur des Proteophosphomannans. Biochem. Physiol. Pflanzen, **175**, 781–788.

Grimmecke, H.-D. und Reuter, G., 1981. Struktur der Zellwandpolysaccharide in der Futtereiweißhefe *Candida spec.* H. VI. Isolierung und Strukturaufklärung der Glucane. Z. Allg. Mikrobiol., **21**, 643–650.

Käppeli, O. and Fiechter, A., 1977. Component from the cell surface of the hydrocarbon-utilizing yeast *Candida tropicalis* with possible relation to hydrocarbon transport. J. Bacteriol., **131**, 917–921.

Lehle, L., Schulz, I. and Tanner, W., 1980. Dolichyl phosphate linked sugars as intermediates in the synthesis of yeast mannoproteins: an *in vivo* study. Arch. Microbiol., **127**, 231–237.

Lehle, L. and Tanner, W., 1978. Biosynthesis and characterization of large dolichyl diphosphate-linked oligosaccharides in *Saccharomyces cerevisiae.* Biochim. biophysica Acta, **539**, 218–229.

Mayer, R. M., 1971. The enzymatic synthesis of the phosphomannan of *Hansenula capsulata.* Biochim. biophysica Acta, **252**, 39–47.

Nakajima, T. and Ballou, C. E., 1974. Characterization of the carbohydrate fragments obtained from *Saccharomyces cerevisiae* mannan by alkaline degradation. J. biol. Chemistry, **249**, 7679–7684.

Nakajima, T. and Ballou, C. E., 1975. Microheterogeneity of the inner core region of yeast manno-protein. Biochem. Biophys. Res. Commun., **66**, 870–879.

Röber, B. und Reuter, G., 1979. Biosynthese der Zellwand-Polysaccharide Mannan und Glucan als Spiegelbild unterschiedlicher Abbauwege der Glucose durch *Candida spec.* H. Z. Allg. Mikrobiol., **19**, 187–194.

Röber, B. und Reuter, G., 1980. GDP-Mannose abhängiger Mannosyl-Transfer in einer Protoplasten-Partikel-Fraktion von *Candida spec.* H. Wiss. Z. d. E.M.A.-Universität Greifswald, Math.-Naturw. Reihe, **29**, 115.

Slodki, M. E., 1962. Phosphate linkages in phosphomannans from yeast. Biochim. biophysica Acta, **57**, 525–533.

Slodki, M. E., 1963. Structure of *Hansenula capsulata* NRRL Y-1842 phosphomannan. Biochim. biophysica Acta, **69**, 96–102.

Thieme, T. R. and Ballou, C. E., 1971. Nature of the phosphodiester linkage of the phosphomannan from the yeast *Kloeckera brevis.* Biochem., **10**, 4121–4129.

Anschrift: Prof. Dr. G. Reuter
Sektion Biologie der Friedrich-Schiller-Universität
— Mikrobielle Biochemie —
DDR 69 Jena, Wöllnitzer Str. 7

Zeitschrift für Allgemeine Mikrobiologie	22	9	1982	675–684

Buchbesprechungen

M. ALEXANDER (Editor), Advances in Microbial Ecology, Vol. 5. XII + 250 S., 17 Abb., 12 Tab. New York—London 1981. Plenum Press. $ 35.00.

Die 5. Folge der Advances bringt 5 gleichermaßen interessante Beiträge. Mangelbedingungen am Standort sind das Thema von 3 Beiträgen: J. L. LOCKWOOD und A. B. FILONOV berichten über das Verhalten von Pilzen bei Mangel an Nährstoffen und Energiequellen an natürlichen Standorten, besonders im Boden und über das Vorkommen von Mykostase und Mykolyse. Oligotrophie, das Leben bei ständiger Nährstoff-Beschränkung auf 1—15 mg organisch gebundenen Kohlenstoff im Liter ist das Thema eines Beitrages von J. S. PAINDEXTER. Stress-Bedingungen durch Wassermangel behandelt D. M. GRIFFIN, ausgehend von den physikalischen Grundlagen. Einen Überblick über Wesen und Vorkommen der vier Mykorrhizatypen (Ektotrophe, Vesikel/Arbuskel-, *Ericales*- und Orchidaceen-Mykorrhiza) geben B. MOSSE, D. F. STIBLEY und F. LE TACON. Im letzten Beitrag berichtet J. G. ZEIKUS über Beziehungen, die zwischen Synthese und Abbau von Lignocellulose und Lignin und dem Kohlenstoff-Kreislauf bestehen.

W. SCHWARTZ (Braunschweig)

H. BOTHE and A. TREBST (Editors), Biology of Inorganic Nitrogen and Sulfur. VII + 384 S., 144 Abb., 34 Tab. Berlin—Heidelberg—New York 1981. Springer-Verlag.

Stickstoff und Schwefel haben durch ihr Auftreten in allen Eiweißen viel Gemeinsames in der Lebewelt. Darüber hinaus ist bei beiden das Vorkommen in unterschiedlichen Oxidationsstufen augenfällig. Die stärker reduzierten Formen (NH_3, NO_2^-, H_2S, S) können Mikroben als Elektronendonatoren für die aerobe Energieerzeugung und die CO_2-Fixierung im Dunkeln dienen, die oxidierten Stufen (NO_3^-, NO_2^-, SO_4^{2-}, SO_3^{2-}) stellen Elektronenakzeptoren für die Energiegewinnung unter anaeroben Verhältnissen dar. Dies war der Anlaß, Stickstoff und Schwefel auf einer internationalen Konferenz in Bochum (BRD) im Mai 1980 vergleichend zu behandeln. Die dort gehaltenen 32 Vorträge sind im vorliegenden Sammelband vereinigt. Ausgehend vom N- und S-Kreislauf in der Natur werden von international anerkannten Wissenschaftlern die einzelnen Prozeßketten näher analysiert. In Übersichts- und z. T. auch experimentellen Beiträgen stehen vor allem die seit etwa 1975 gewonnenen Erkenntnisse kritisch zur Diskussion, wobei teilweise ganz neue Entwicklungslinien sichtbar werden. Beeindruckend sind vor allem die großen Fortschritte bei der Aufklärung der mikrobiellen N_2-Fixierung durch kombinierte Anwendung molekulargenetischer, biochemischer und physiologischer Methoden. Sie dürften ein Beispiel dafür sein, wie auch auf anderen Gebieten die Tür zum Fortschritt zu öffnen ist. Der Band stellt eine Fundgrube neuer Erkenntnisse für alle auf einschlägigen Gebieten wissenschaftlich Arbeitenden dar und hebt sich vom Niveau vieler anderer Kongreßberichte positiv ab.

G. SCHILLING (Halle)

F. E. CLARK and T. ROSSWALL (Editors), Terrestrial Nitrogen Cycles, Processes, Ecosystem, Strategies and Management Impacts (Ecological Bulletins No. 33). 709 S., 167 Abb., 121 Tab. Stockholm 1981. Swedish Natural Science Research Council. DM 116,00.

Einer der die Pflanzenproduktion limitierenden Faktoren ist der Stickstoff. Bei der derzeitigen Düngungspraxis wird jedoch etwa die Hälfte des zugeführten Stickstoffs nicht von den Pflanzen verwertet, gleichzeitig führt der nicht verwertete Anteil zu Umweltbelastungen. Einen Beitrag zur Lösung dieser aktuellen Problematik strebt das vorliegende Buch an. Es geht auf einen 1979 in Schweden durchgeführten internationalen Workshop zurück, bei dem kompetente Mikrobiologen, Ökologen sowie Land- und Forstwissenschaftler zur interdisziplinären Beratung zusammenkamen. Drei Problemkreise wurden behandelt, die am Stickstoffkreislauf beteiligten Prozesse, die Strategien der Stickstoffrezirkulation in natürlichen Ökosystemen und Aspekte des land- und forstwirtschaftlichen Managements. Unter den Prozessen werden sowohl mikrobielle Umsetzungen (Abbau organischer Substanz, Nitrifikation und Denitrifikation) als auch Mikroben-Pflanzen-Interaktionen bei der Stickstoffbindung und Mykorrhiza behandelt. Der Bogen der behandelten Ökosystem-Strategien reicht von Tundren über Wiesen und Wälder gemäßigter

Zonen bis zu den Steppen und Regenwäldern der Tropen. Bei den landwirtschaftlichen Problemen finden neben der Düngung die Kontrolle der Nitrifikation, Beweidung sowie Be- und Entwässerung Berücksichtigung. Den drei Problemkreisen sind allgemeine Beiträge vorangestellt, unter denen ein quantitatives Modell der Kohlenstoff- und Stickstoff-Dynamik zwischen Pflanzen, Mikroorganismen und Boden von besonderem Interesse ist. Mit einer zusammenfassenden Betrachtung über den Erkenntnisstand und daraus abgeleitete Aufgaben für die weitere Forschung endet der Band. Durch die Konzeption des Workshops, das der Buchpublikation zugrunde liegt, wurde eine Geschlossenheit und Qualität der insgesamt 50 Beiträge erreicht, die weit über die vieler Tagungspublikationen hinausgeht. Es wird deutlich, daß zur Steigerung der landwirtschaftlichen Produktion bei gleichzeitiger Vermeidung oder Minimierung von Umweltbelastungen eine wesentlich komplexere Betrachtungsweise und Forschungsstrategie als bisher üblich notwendig ist. Für die Mikrobiologen zeichnet sich die große Aufgabe ab, in Zusammenarbeit mit anderen Wissenschaftsdisziplinen Wege zur Optimierung mikrobieller Leistungen im Boden zu finden. Natürliche Ökosysteme, die eine hohe Effektivität der Stickstoffverwertung zeigen, z. B. geringe Nitrifikationsraten, bieten Ansätze dafür. Den Herausgebern gebührt Dank für die Publikation, die eine Fülle von Informationen und wertvolle Anregungen zur Lösung der eingangs skizzierten Problematik liefert. W. FRITSCHE (Jena)

Current Topics in Microbiology and Immunology, Vol. 94/95. IV + 308 S., 46 Abb., 35 Tab. Berlin-Heidelberg-New York 1981. Springer-Verlag. DM 132,00.

Der Doppelband enthält wieder ein breites Themenspektrum. Von C. W. WARD werden die bekannten Daten über die Struktur des Influenzavirus-Hämagglutinins zusammengestellt und daraus Hypothesen über Antigenshift und -drift abgeleitet. Bei der Antigendrift wiederholt sich die Veränderung eines Aminosäurerests nicht, bei weiteren Mutationen werden andere Aminosäuren ausgetauscht. Bei der Antigenshift werden andere Mechanismen wirksam; für das Auftreten der Hongkong- und Asia-Viren wird ein genetisches Rearrangement angenommen. Umgekehrt wird vermutet, daß die HO/HI-Influenza des Jahres 1977 in gefrorenem Zustand in der Natur (oder sonstwo) überlebte, da ein genetisch gleiches Virus 1950 verbreitet war. H. G. BOMAN und H. STEINER legen Experimente über die Induktion einer Immunität bei Puppen von *Hyalophora cecropia* vor; nach Injektion von *Enterobacter cloacae* kommt es zum Auftreten ziemlich breit, aber nicht unspezifisch wirkender antibakterieller Proteine. — Den Bakteriengenetiker wird besonders der Beitrag von G. HOBOM überReplikationssignale in Prokaryoten-DNA interessieren. — Ein weiteres Virusthema behandeln W. OSTERTAG und I. B. PRAGNELL: Differenzierung und Virusbeteiligung bei der Differenzierung transformierter erythroider Zellen von Mäusen und Ratten. — Von methodischem Interesse ist die Übersicht von J. MEYER über die Elektronenmikroskopie von Virus-RNA. — Der inhaltsreiche Band wird mit der Übersicht von HOCHSTADT *et al.*: Genetische Alteration animaler Zellen in der Kultur, abgeschlossen. W. KÖHLER (Jena)

Current Topics in Microbiology and Immunology, Vol. 96. VIII + 264 S., 63 Abb., 37 Tab. Berlin-Heidelberg-New York 1982. Springer-Verlag. DM 92,00.

Dieser von P. H. HOFSCHNEIDER und W. GOEBEL editierte Band der Schriftenreihe ist mit 13 Beiträgen einem Thema gewidmet: Genklonierung in andere Organismen als *E. coli* (Gene Cloning in Organisms other than *E. coli*). Diese Frage gewinnt immer mehr an Bedeutung, nachdem die grundlegenden Erkenntnisse an *E. coli* erarbeitet wurden. Die in Aussicht genommene industrielle Nutzung zwingt zum Einsatz anderer Organismen, von denen u. a. erwartet wird, daß sie bessere Möglichkeiten der Massenkultur bieten, daß glykosylierte tierische Proteine exprimiert werden usw. In den Beiträgen sind, jeweils von 2 Autoren um Vollständigkeit zu erreichen, die gegenwärtigen Kenntnisse über die Wirtssysteme Bazillen (besonders *B. subtilis*), *Pseudomonas*, *Streptomyces*, *Neurospora crassa* und Hefen zusammengestellt. Ferner wird der Gentransfer in Mammalierzellen behandelt sowie die Übertragung auf pflanzliche Zellen. Ein Beitrag ist der Verwendung von Liposomen zur Übertragung von DNA oder RNA auf Tier-, und Pflanzen-Zellen gewidmet. — Das Buch gibt eine vorzügliche Übersicht des raschen Fortschritts in den letzten Jahren, Literatur ist bis 1981 erfaßt. W. KÖHLER (Jena)

Current Topics in Microbiology and Immunology, Vol. 97. IV + 204 S., 28 Abb., 14 Tab. Berlin-Heidelberg-New York 1982. Springer-Verlag. DM 94,00.

Dieser Band der wohlbekannten Schriftenreihe umfaßt vorwiegend virologische Übersichten. M. R. MACNAUGHTON referiert über die gegenwärtigen Kenntnisse zur Struktur und Replikation von Rhinoviren, die derzeit 115 humane und 2 bovine Typen umfassen, die jedoch weitgehend ähnliche molekulare Strukturen aufweisen. Von der Replikationsstrategie der Rhinoviren ist weitaus weniger bekannt als bei anderen Picornaviren. — In seinem Beitrag über Poxvirus-DNA behandelt J. A. HOLOWCZAK in erster Linie das Poxvirusgenom. H. PERSSON und L. PHILIPSON stellen die gegenwärtigen Kenntnisse zur Regulation der Adenovirus-Genexpression zusammen, einer Virusgruppe, die in den letzten Jahren wegen ihrer relativ geringen klinischen Bedeutung etwas in Vergessenheit geriet. — Von allgemein biologischem Interesse ist die Übersicht von K. H. NIERHAUS über Struktur, Assemblierung und Funktion von Ribosomen, wobei vor allem den Fragen der Struktur, der Zusammensetzung, der Biosynthese und der Funktion nachgegangen wird. Ein interessantes Kapitel darin ist die Untersuchung der Ribosomen 5 S-RNA und des Ribosomenproteins für die Beantwortung von Fragen der Evolution. Der mit 5 S-RNA-Analyse gewonnene Evolutionsbaum gleicht fast völlig den Ergebnissen, die mit der Analyse von Cytochrom C-Sequenzen gewonnen wurden (das aber bei Prokaryoten fehlt). Auf Grund der Ribosomenanalysscheint die Benennung der Archaebakterien als ein drittes Reich neben Eubakterien und Eukaryoeten als falsch. W. KÖHLER (Jena)

Current Topics in Microbiology and Immunology, Vol. 98: E. WECKER and I. HORAK (Editors) Retrovirus Genes in Lymphocyte Function and Growth. VIII + 142 S., 8 Abb., 11 Tab. Berlin-Heidelberg-New York 1982. Springer -Verlag. $ 30.80.

Der von E. WECKER und J. HORAK editierte Band der Reihe ist einem Thema gewidmet, der Bedeutung von Retrovirusgenen für Funktion und Wachstum von Lymphocyten und enthält 10 Beiträge von 17 Autoren. Die Mehrzahl davon wurde bei einem Symposium in Würzburg 1981 vorgestellt, liegt aber hier in einer speziell überarbeiteten, die Thematik weiter fassenden Form vor. Zwei große Themengruppen werden behandelt: Die Klassen endogener Viren und ihr Ursprung und die Struktur der integrierten Retrovirusgenome und ihre biologischen Wirkungen, COUTINHO legt eine interessante Hypothese über den Einfluß von Retroviren auf die Evolution des Immunsystems der Vertebraten vor. Diese Viren können das Wachstum von T-Lymphocyten ohne Mitwirkung akzessorischer Zellen oder ihrer Produkte stimulieren; sie wirken damit als „echte T-Zellen-Mitogene". Viele Probleme sind noch ungeklärt; das Buch gibt Hinweise auf offene Fragen und Anstöße für weiterführende Arbeiten. W. KÖHLER (Jena)

I. C. FELKNER (Editor), Microbial Testers: Probing Carcinogenesis (Microbiology Series, Vol. 5). 288 S., 38 Abb., 32 Tab. New York 1981. Marcel Dekker, Inc. SFr 95.00.

Zur Prüfung von Chemikalien auf mutagene und karzinogene Wirkungen sind aussagefähige Tests erforderlich. Bei der großen Zahl der zu testenden Verbindungen müssen sie einen hohen Durchsatz ermöglichen sowie schnell und ohne großen Aufwand zu sicheren Ergebnissen führen. Mikrobielle Mutagenitätstests haben sich dafür als sehr geeignet erwiesen, der von AMES entwickelte Test mit *Salmonella*-Mutanten in Kombination mit tierischen Mikrosomenfraktionen findet als Vortest breite Anwendung. Jedoch haben diese Tests auch ihre Grenzen, u. a. werden nicht alle karcinogenen Chemikalien erfaßt. Das Anliegen der vorliegenden Buches ist es, zur weiteren Entwicklung mikrobieller Tests beizutragen. Durch die Zusammenarbeit von Vertretern verschiedener Wissenschaftsdisziplinen, wie der Mikrobiologie, Genetik, Chemie, Toxikologie und Medizin, ist es dem Herausgeber gelungen, ein umfassendes wie tiefgründiges Bild vom Stand und den Tendenzen der Testproblematik zu vermitteln. Für den Mikrobiologen sind unter den zehn Beiträgen vier von besonderem Interesse, die sich mit der weiteren Entwicklung spezifischer und rationeller Tests befassen. Dabei werden neben dem eingangs erwähnten Kombinationstest mit gramnegativen Bakterien Testsysteme mit *Bacillus subtilis* (u. a. DNS-reparatur-defekte Mutanten) und *Neurospora* (Adenin-Mutanten) behandelt. Durch die Berücksichtigung tierischer Testsysteme wird die Stellung mikrobieller Systeme in der Testhierarchie deutlich. In zwei abschließenden Beiträgen zur multifaktoriell bedingten Karzinogenese, bei der die Ernährungsweise und die Intestinalflora eine Rolle spielen, wird die Notwendigkeit komplexerer Tests erläutert. Mit einem Beispiel über den Einfluß von Metaboliten des Fettstoffwechsels auf die chemische Karzinogenese werden die Möglichkeiten komplexer mikrobieller Tests belegt. Das Buch führt das große mikrobielle Potential für neue hochwirksame Mutagenität- und Karzinogenitätstests vor Augen. Diese Informationen geben wertvolle Anregungen für andere Anwendungsgebiete von mikrobiellen Tests. Das Buch demonstriert, wie durch interdisziplinäre Zusammenarbeit ein aktuelles Problem des vorbeugenden Gesundheits- und Umweltschutzes einer weiteren Lösung zugeführt wird. W. FRITSCHE (Jena)

B. Lundholm and M. Stacker (Editor), Environmental Protection and Biological Forms of Control of Test Organisms (Ecological Bulletins No. 31, Proceedings of an International Workshop arranged by Swedish Comission for Research on Natural Resources and the Swedish Products Control Board, Stockholm 14—17 May, 1979). 171 S., 9 Abb., 14 Tab. Stockholm 1980. Swedish Natural Science Research Council. $ 25.00.

Die vorliegende Publikation ist der Frage gewidmet, ob bei Mitteln zur biologischen Schädlingskontrolle umweltrelevante und toxikologische Neben- und Spätwirkungen auftreten. Die Herausgeber führten dazu 1979 in Stockholm einen Workshop durch, zu dem sie kompetente Wissenschaftler und Vertreter der für die Zulassung verantwortlichen staatlichen Institutionen aus 9 Ländern zusammenführten. Die 12 Beiträge und eine Gesamteinschätzung umfassende Publikation basiert auf diesem Workshop. Das Gebiet der biologischen Kontrolle ist weit gefaßt, es berücksichtigt neben Parasiten und Räubern auch Pheromone, Insektenhormone, biogene Wirkstoffe und die Resistenzzüchtung. Für den Mikrobiologen sind die Beiträge über mikrobielle Insekticide auf der Basis von insektenpathogenen Viren, Bakterien und Pilzen von besonderem Interesse. Eine Reihe von Bakterien- und Viruspräparaten haben sich als unbedenklich erwiesen. *Bacillus thuringiensis* und *B. popilleae* sowie Kernpolyedervirus-Präparate sind in einigen Ländern zugelassen. Pilzpräparate befinden sich dagegen noch in der Überprüfung. Auch einige Antagonisten von phytopathogenen Mikroorganismen sind bereits zugelassen, z. B. *Agrobacterium radiobacter*. In drei Beiträgen werden die in den USA, Großbritannien und Frankreich erforderlichen Maßnahmen für eine Zulassung von mikrobiellen Pesticiden behandelt. Mit den klaren Aussagen über die Risikoeinschätzung und den Informationen und Empfehlungen über Sicherheitstests leistet der Band einen wertvollen Beitrag zur weiteren Entwicklung von Mitteln und Methoden der biologischen Schädlingskontrolle. W. Fritsche (Jena)

S. Mitsuhashi (Editor), Beta-Lactam Antibiotics. XII + 325 S. Berlin—Heidelberg—New York 1981. Springer-Verlag. DM 98,00.

Der Band beginnt mit einer kurzen Geschichte der β-Lactam-Antibiotica, führt zur Epidemiologie und zu den genetischen Grundlagen der Resistenz von Mikroorganismen gegen β-Lactam-Strukturen und charakterisiert in einem eigenen Kapitel die biochemischen Mechanismen, die der β-Lactam-Resistenz zugrunde liegen. In folgenden Kapiteln werden Struktur-Wirkungsbeziehungen offenbart, die insbesondere bei neuentwickelten Verbindungen wie Ampicillinen, Apalcillin und verwandten Strukturen, aber auch bei Penicillinen und Cephalosporinen mit der 2,3-Dioxopiperazin-Gruppe, ferner bei Cefsulodin, Cefotiam, Cefmenoxim, Cefotaxim, Ceftizoxim und Cefmetazol sowie bei Cephamycin-Derivaten erkannt worden sind. Besonders interessant nehmen sich die Vertreter einer neuen Familie von β-Lactam-Strukturen aus, die als sog. Olivanat-Familie (auch PS-Serie) auf Grund ihrer antimikrobiellen Aktivität bei gleichzeitiger Inhibitorwirkung gegen β-Lactamasen bekannt geworden sind. Zu dieser Gruppe von Verbindungen gehören die Thienamycine, Olivansäuren und PS-Verbindungen, die alle aus Streptomyceten isoliert werden können. In Kapiteln über die Wirkungsweise der β-Lactam-Antibiotica in Mikroorganismen werden besonders die Funktionen der Penicillin-bindenden Proteine 5 und 6 bei *Escherichia coli* erläutert, die Eigenschaften der DD-Carboxypeptidasen und die Rolle der Penicillin-bindenden Proteine bei der Zellteilung und Sensibilität gegenüber β-Lactamen. In weiteren Kapiteln wird über die Pharmakokinetik der Penicilline und Cephalosporine sowie über die Allergien ausgeführt, die nach Behandlung mit β-Lactam-Antibiotica bei Patienten auftreten. Insgesamt liegt ein sehr gut ausgestattetes Buch vor, dessen Wert für den Wirkstoff-Forscher unbestritten ist. W. F. Fleck (Jena)

S. Mitsuhashi, L. Rosival and V. Krĉmery (Editors), Antibiotic Resistance — Transposition and other Mechanisms (Forth International Symposium on Antibiotic Resistance, Castle of Smolenice, Czechoslovakia, 1979). XII + 410 S., 124 Abb., 137 Tab. Prague, Berlin-Heidelberg-New York 1980. Avicenum, Czechoslovak Medical Press, Springer-Verlag.

Das Symposium „Antibiotic Resistance", dessen 60 Vorträge von Autoren aus über 20 europäischen Ländern in dem vorliegenden Berichtsband in vollem Wortlaut abgedruckt sind, ist das vierte in einer Reihe von international renomierten wissenschaftlichen Veranstaltungen. Die zur Diskussion gestellten Beiträge geben ein repräsentatives Bild des Standes und der Entwicklung der Forschung auf dem Gebiet der Antibiotica-Resistenz. Gewährleistet wird dies durch ein ausgewogenes Verhältnis von Übersichtsreferaten und Originalbeiträgen. Wegen der globalen Bedeutung der Antibiotica-Resistenz wurde das Symposium in zwei Abschnitte geteilt. Der erste ist vorwiegend den theoretischen Aspekten gewidmet und behandelt die genetischen und molekularbiologischen Grundlagen der R-Plasmide und der Transposons. Der zweite Abschnitt hingegen ist mehr praxisorientiert, wobei die medizinische Bedeutung der Antibiotica-Resistenz im Vordergrund steht. Hierbei nehmen neben Antibiotica-Strukturen und Resistenzmechanismen die ökologischen und epidemiologischen Prozesse einen breiten Raum ein. Sehr zu begrüßen ist neben dem Autorenverzeichnis das umfangreiche Stichwortregister. Die ausgezeichnete drucktechnische Ausführung gestattet auch die Wiedergabe von Halbtonbildern. Das Buch wird deshalb nicht nur für Fachleute der molekularbiologischen und medizinischen Forschung nützlich sein können, sondern auch in der postgradualen Ausbildung eine reale Hilfe sein. D. NOACK (Jena)

A. MOSER, Bioprozeßtechnik — Berechnungsgrundlagen der Reaktionstechnik biokatalytischer Prozesse XV + 198 S., 95 Abb., 13 Tab. Wien-New York 1981. Springer-Verlag. DM 69,00.

Dieses Lehrbuch ist Ergebnis einer mehr als zehnjährigen Tätigkeit des Autorsin Forschung und Lehre auf dem Gebiet der Biotechnologie an der Technischen Universität Graz (Österreich) sowie als Gastprofessor im Ausland. Es ist dem Autor eine geschlossene und didaktisch bestechend klare Darstellung dieses noch jungen und recht heterogenen Wissensgebietes gelungen. Das Schwergewicht des Inhaltes ist im Untertitel treffend charakterisiert: Anwendung von Ergebnissen der physikalischen, mikrobiologischen und biochemischen Grundlagenforschung auf die quantitative Analyse sowie den modellgestützten Entwurf von technisch-mikrobiologischen Prozessen und Fermentoren.

Das Buch ist nach einer vorangestellten ausführlichen Nomenklaturliste wie folgt gegliedert: 1. Einführung (10 S.), 2. Die Bioprozeßtechnik und ihre Arbeitsprinzipien (42 S.), 3. Bioreaktoren (45 S), 4. Prozeßkinetische Analyse (31 S.), 5. Formalkinetik von Bioprozessen (41 S.), 6. Prozeßentwurf: Methoden der Voraussage des Umsatzes bzw. der Produktivität (30 S.). Bei der Kürze der Darstellung im Rahmen dieses Lehrbuches wird jeder Wissenschaftler und Ingenieur auf seinem eigenen Spezialgebiet bestimmte Aussagen vereinfacht wiederfinden oder vermissen, jedoch wird ihm diese Kürze bei der Einarbeitung in noch fremde, im Lehrbuch übersichtlich dargebotene Disziplinen sehr hilfreich sein. Auf Grund sehr umfangreicher Literaturhinweise wird der Zugang zur weiteren Erschließung der einzelnen Gebiete ermöglicht. Das Lehrbuch stellt die inhaltlichen Zusammenhänge zwischen den einzelnen Gebieten her und führt insbesondere die mikrobiologisch-biochemisch-biophysikalische, auf Grundlagenforschung gerichtete Denkweise mit der biotechnologischen ingenieurmäßig auf den Prozeßentwurf gerichteten Denkweise fruchtbar zusammen. ,,Nach Erarbeitung der qualitativen Eigenschaften des Systems Biokatalysator/Substrat im mikrobiologischen Labor bildet die quantitative Erfassung und mathematische Modellierung der Kinetik der Bioprozesse im 'Perfekt-Bioreaktor' die erste Arbeitsphase (vgl. Kapitel 4, 5), die Quantifizierung und Modellierung der verschiedenen Bioreaktoren im 'Bioreaktormodell' die zweite Phase (Kapitel 3), während der Prozeßentwurf (Kapitel 6) erst abgeschlossen ist, wenn ein Prozeßmodell aus Kinetik und Transportvorgängen (und ein scale-up-Kriterium) aus zusätzlichen Experimenten in Pilot-Anlagen erarbeitet wurde.“

Das Buch ist mit graphischen Darstellungen sehr reich illustriert.

Es wird einen Gewinn bedeuten für jeden Leser, der sich in neue, kinetisch orientierte biotechnologische Teildisziplinen einarbeiten muß, einerlei ob er Student, Wissenschaftler oder Ingenieur ist.

R. GUTHKE (Jena)

W. NAGL, Elektronenmikroskopische Laborpraxis. 143 S., 26 Abb., 42 Taf. Berlin-Heidelberg-New York 1981. Springer-Verlag.

Bei der vorliegenden ,,Elektronenoptischen Laborpraxis“ handelt es sich, wie im Untertitel auch angegeben, um eine ,,Methodensammlung mit Bildbeispielen für Labor und Forschung in der Medizin und Zellbiologie“. Neben einer Darstellung ausgewählter, jedoch häufig eingesetzter Routinemethoden in der Elektronenmikroskopie werden sowohl theoretische Grundlagen zur Mikroskopie, zum Fixierungs- und Einbettungsprozeß in Kunstharze vermittelt wie auch Einführungen in die Aufnahmetechnik von Elektronenmikrogrammen einschließlich ihrer Bewertung und in weitere elektronenoptische Techniken (Abdruckverfahreu, SEM, STEM, Elementanalyse) vorgenommen. Die Methodensammlung reicht im Gebiet der Transmissionselektronenmikroskopie von der chemischen und physikalischen Fixierung über die Kunstharzeinbettung, die Ultramikrotomie bis zur Kontrastierung und einen Abriß spezieller Techniken, wie gerichtete Einbettung, Flacheinbettung, Serienschnittechnik und Rekonstruktion, Ultracytochemie und Autoradiographie, die Spreittechnik und Gefrierätzung. Die Wiedergabe bewährter Verfahrensvorschriften, das Aufzeigen kritischer Punkte und der Möglichkeiten zur Überwindung auftretender Schwierigkeiten, die knappe und übersichtliche Darstellung der Sachverhalte verbunden mit einer Beschränkung auf Wesentliches und Prinzipielles sowie nicht zuletzt die auch für den wenig Versierten leicht verständliche Schreibweise lassen diese „Laborpraxis" zu einer gelungenen Bereicherung der einführenden elektronenoptischen Literatur werden. Ein Verzeichnis vertiefender Literatur, ein Bezugsquellennachweis, ein ausführliches Register und Bildtafeln zur Ultrastruktur der Zelle steigern den Gebrauchswert beträchtlich. Für Studenten, aber auch für technische Assistenten und Wissenschaftler, die den Einstieg in die Praxis biologisch-medizinischer Elektronenmikroskopie suchen, stellt diese von NAGL vorgelegte Methodensammlung ein ausgezeichnetes Mittel dar. E. UNGER (Jena)

H. J. PREUSSER, *Trichophyton rubrum, Candida albicans*, Biologie, Cytologie, Wirkung von Imidazol-Präparaten. IX + 109 S., 83, zum Teil farbige Abb., 7 Tab. Stuttgart—New York 1982. Gustav Fischer Stuttgart—New York. DM 32,00.

Beide Pilzarten sind in Mitteleuropa weit verbreitet als Erreger von Dermatomykosen; *C. albicans* kann auch Schleimhaut-Erkrankungen und systemische Mykosen verursachen. Beide Arten werden in bezug auf Kolonieform, Zelltypen, Wachstum, elektronenmikroskopische Zellstruktur beschrieben. Davon ausgehend wird untersucht, welche Veränderungen das antibiotisch wirkende Imidazol-Präparat Econazol in der Feinstruktur der Zellen bewirkt und wie Wachstum und Überlebensrate beeinflußt werden. Die reichlichen zum Teil farbigen Abbildungen zeigen Fälle von *Trichophyton*- und *Candida*-Mykosen beim Menschen, Kolonieformen auf künstlichen Substraten, Aufnahmen der Feinstruktur normaler und unter Antibioticum-Einfluß gewachsener Zellen. Die besonders in der Cytoplasmamembran auftretenden Veränderungen werden diskutiert, desgleichen Fragen der klinischen Anwendung von Econazol. Weiterführende Literatur ist angegeben. W. SCHWARTZ (Braunschweig)

H. J. REHM and G. RIED (Editors), Biotechnology, a Comprehensive Treatise in 8 Volumes. Vol. 1: Microbial Fundamentals. 529 S., 237 Abb., 104 Tab. Weinheim—Deerfield Beach, Fla.—Basel 1981. Verlag Chemie. DM 495,00.

In Band 1 der geplanten 8bändigen Biotechnologie in englischer Sprache werden von 16 Autoren die mikrobiologischen Grundlagen und Voraussetzungen biotechnischer Verfahren behandelt. Band 2 wird sich auf technische Möglichkeiten des Arbeitens mit Mikroorganismen beziehen, die Bände 3 und 4 werden die Erzeugung von Biomasse und die Produktion von Metaboliten des primären und sekundären Stoffwechsels, wie organische Säuren, Alkohole, Antibiotica, Alkaloide, Psychopharmaca behandeln, Band 5 hat als Thema die mikrobiologische Bereitung alkoholischer Getränke, von Molkereiprodukten, von sonstigen Lebensmitteln und von Futtermitteln. Band 6 wird sich auf mikrobiologisch-biochemische Transformierung von Steroiden und auf Gewebekultur von Pflanzen und Tieren, auf Stickstoffbindung, auf Leaching beziehen, Band 7 auf Gewinnung und Möglichkeiten der Verwendung von Enzymen, und Band 8 wird die mikrobiologische Aufbereitung von Abwasser und von organischen Abfällen sowie den mikrobiellen Abbau spezieller organischer Verbindungen behandeln, — eine großzügig angelegte Planung.

Die Microbial Fundamentals des ersten Bandes sind in 7 Kapitel aufgeteilt. REHM und REED leiten das Ganze ein mit einer Begriffsbestimmung der Biotechnologie, aus der nach einer Definition der European Federation of Biotechnology landwirtschaftlich-mikrobiologische und medizinische Technologie ausgeschlossen sind. Im ersten Kapitel (67 Seiten) charakterisiert ANNA KOCKOVA-KRATOCHYLOVA industrielle Mikroorganismen und Viren. CH. L. COONEY behandelt im zweiten Kapitel Methoden zur Bestimmung der Vermehrung von Mikroorganismen in Massenkultur (40 S.). Das umfangreiche dritte Kapitel (98 S.) von H. W. DOELLE ist biochemisch orientiert und gibt eine Übersicht über Membrantransport, aerobe und anaerobe Stoffwechselwege, Schwefel- und Stickstoff-Kreislauf. Gegenstand von Kapitel 4 ist eine Darstellung der Synthesewege sekundärer Metabolite bei industriell genutzten Mikroorganismen von J. F. MARTIN und P. LIRAS (23 Seiten). Die Beziehungen zur Genetik werden in Kapitel 5 dargestellt (118 S.), aufgeteilt in eine Übersicht über die Anwendbarkeit genetischer Möglichkeiten (F. P. ELANDER und A. L. DEMAIN) und in die Abschnitte Mutationen (G. K. JACOBSON), Hybridisation (K. ESSER und U. STAHL), genetische Technik (A. PÜHLER und W. HEUMANN). Die Frage nach der Herkunft industriell genutzter Mikroorganismen und nach den Möglichkeiten eines Screening, z. B. für das Auffinden antibiotisch wirksamer Stämme (K. NAKAYAMA) und die Methoden der Reinkultur, angewandt auf industriell genutzte Mikroorganismen (A. DIETZ, 19 S.) sowie patentrechtliche Fragen (V. VOSSIUS, 17 S.) sind Gegenstand des 6. Kapitels (98 S.). Das 7. Kapitel (A. FIECHTER, 53 S.) geht auf Fragen der Massenkultur und der kontinuierlichen Kultur von Mikroorganismen sowie von Pflanzen- und Tierzellen ein. — Die Abbildungen machen den Text anschaulich; sie bringen z. B. schematische Darstellungen von Stoffwechselwegen und zur Anwendung genetischer Verfahren, Diagramme zur Chemostatenkultur. Bis Ende 1984 sollen nach der Planung die 8 Bände erschienen sein. Anlage und Umfang des Ganzen lassen die zunehmende Bedeutung mikrobiologisch-technischer Verfahren erkennen und entsprechen dem enzyklopädischen Zug unserer Zeit.

W. SCHWARTZ (Braunschweig)

H. REICHENBACH and O. B. WEEKS (Editors), The *Flavobacterium-Cytophaga* Group (Proceedings of the International Symposium on Yellow-Pigmented Gram-Negative Bacteria of the *Flavobacterium-Cytophaga* Group, 8 to 11 July 1980, Braunschweig-Stöckheim). 217 S., 7 Abb., 107 Tab. Weinheim—Deerfield Beach, Florida—Basel 1981. Verlag Chemie. DM 88,00.

Der Symposiumsband enthält 23 Beiträge zur Geschichte, Nomenklatur, Taxonomie, Biochemie (% GC-Werte der DNS, DNS:RNS-Hybridisation, Fettsäurespektren, Strukturaufklärung der Pigmente), Ökologie und klinischen Bedeutung der *Flavobacterium-Cytophaga* Gruppe.

Auf einen kurzen, einfachen Nenner gebracht, stehen beide Gattungsnamen für gelb- bzw. orange-pigmentierte, gram-negative Bakterien aus natürlichen und anthropogenen Biotopen. Sie besitzen eine nicht zu unterschätzende praktische Bedeutung, da sie relativ häufig in klinischem Untersuchungsmaterial auftreten und als Ursachen von Nahrungsmittelverderb in der Industrie, als Erreger von Fischkrankheiten und als Produzenten interessanter Enzyme und Antibiotica eine Rolle spielen. Schließlich wird der Ökologe mit ihnen konfrontiert, da sie in großen Populationen im Boden, im Süßwasser und marinen Gewässern vorkommen und zweifellos eine Rolle beim Abbau von organischen Substraten in der Natur und in Abwasseranlagen spielen.

Beide Gattungen umfassen stäbchenförmige, heterotrophe, aerobe Bakterien, die sich dadurch unterscheiden, daß *Cytophaga* durch sog. „gliding movement" ausgezeichnet ist, während *Flavobacterium* beweglich ist mit Hilfe von Geißeln oder unbeweglich. Das Problem besteht nun darin, daß 1. die Feststellung der Art und Weise der Fortbewegung zeitraubend und nicht immer eindeutig möglich ist, daß es 2. gewisse Überschneidungen zwischen beiden Gattungen (sog. „*Cytophaga*-like bacteria") bezüglich der Morphologie und auch bezüglich des Pigmenttyps gibt und daß 3. in beiden Gattungen Taxa mit niedrigen und hohen GC-Werten unterschieden werden können. Besonders aus der letztgenannten Tatsache muß geschlossen werden, daß beide Gattungen in der gegenwärtigen Fassung heterogen sind.

Die Beiträge des Symposiums bilden gewissermaßen die Bestandsaufnahme aller Erkenntnisse der letzten 15 Jahre seit das *Flavobacterium-Cytophaga*-Problem durch die Aufdeckung stark divergierender GC-Werte in beiden Gattungen in Bewegung geraten ist und — wie diese Beiträge beweisen — vorläufig noch weit entfernt ist von einer befriedigenden Lösung.

Als eine begrüßenswerte Initiative des Symposiums ist der Vorschlag zu werten, sog. „temporary collections" einzurichten, die für eine gewisse Zeit diejenigen Stämme halten sollten, die bestimmte Typen, Phena oder Cluster repräsentieren. Das könnten auch noch nicht benannte oder vorläufige Phena oder Taxa sein. Damit würden zukünftige Bearbeiter in die Lage versetzt, genau dort fortzufahren, wo frühere aufgehört haben, bzw. würden Ergebnisse, die mittels unterschiedlicher Methoden gewonnen wurden, untereinander vergleichbar. Die NCTC, London (für *Flavobacterium*) und die DSM, Göttingen (für *Cytophaga*-ähnliche Bakterien) wurden als „Aufenthaltsorte" für derartige „temporary collections" vorgesehen.

JUTTA MEYER (Jena)

D. R. REYNOLDS (Editor), Ascomycete Systematics — The Luttrellian Concept. IX + 242 S., 122 Abb. Berlin-Heidelberg-New York 1981. Springer-Verlag. DM 79,00.

Themen des 2. Internationalen Mycologie-Kongresses (1977) wurden als erweiterte Komplexe in Hinsicht auf das von LUTTRELL 1951 und 1955 vorgestellte Konzept zur Ascomyceten-Systematik zusammengestellt. Im ersten Abschnitt des Buches berichtet BECKETT über den Ascus mit Apikalporus (20 S.) und bezieht Entwicklung, Aufbau (insbesondere Ascus-Wand und Apikalapparat) und Funktion in die Betrachtung ein. Über 22 Seiten behandelt VAN BRUMMELEN den operculaten Ascus und die verwandten Formen. Dabei wird von 45 Species der Aufspringmechanismus (Licht- und Elmi-Untersuchungen) besonders berücksichtigt und die Ableitungsverhältnisse herausgearbeitet. MÜLLER berichtet dann (5 S.) über den bitunicaten Ascus, wobei auf den nachweisbaren vielschichtigen Aufbau der Wände und die Verwendung des Begriffes „bitunicat" eingegangen wird. BELLEMERE und LETROUT-GALINOU behandeln anschließend (16 S.) die Ultrastruktur des Lecanoralen-Ascus, wobei anhand zahlreicher Abbildungen und Skizzen auf die vier Haupttypen (*Parmelia*, *Placynthium*, *Collema* und *Cladonia*) ausführlich eingegangen wird. Der zweite Abschnitt des Buches behandelt dann Ascocarp-Komponenten, die mit ihren zentralen Anteilen des Primordiums (ascocarpbud) als Centrum (carpocenter) bezeichnet werden.

Die schon länger umstrittene Gruppierungs-Bezeichnung der „Plectomycetes" wird bezüglich des Centrums von MALLOCK untersucht (19 S.). Dabei werden Species der *Pleosporales*, *Hypocreales*, *Diaporthales* und *Pezizales* einbezogen und die Heterogenität der „Ordnung" *Plectascales* (auch was ursprüngliche und abgeleitete Species sind) unterstrichen. KIMBROUGH setzt sich mit dem Discomycetes-Centrum auseinander (10 S.). Die Bildung des Ascogoniums bei den inoperculaten und operculaten Formen und die Entstehung des Hymeniums werden herausgestellt. PARGUEY-LEDUC und JANEX-FAVRE behandeln anschließend (22 S.) die Ascocarpe der ascohymenialen Pyrenomyceten. Unterstrichen wird, daß für die Systematisierung neben dem Fruchtkörper-Aufbau der Ascus sowie Ascosporenformen, die Ascusstruktur und die Ascocarp-Entwicklung immer wichtiger werden. Behandelt werden deshalb das Ascothecium, Carpocenter, Paraphysen, Ostiolar-Apparat neben den Ascocarp-Typen und die Entwicklung der Carpocenter-Typen von *Diaporthe*, *Xylaria*, *Eutypa* und Species der *Pyrenolichenes*. LUTTRELL berichtet anschließend über das Centrum der Loculoascomycetes (15 S.). Unter Beifügung zahlreicher Abbildungen wird der Ursprung der Ascocarpe und deren Entwicklungsmuster untersucht. HENSSEN in Kooperation mit KEUCK, RENNER und VOBIS behandeln abschließend in einer sehr umfangreichen Studie (96 S.) das Centrum der *Lecanorales*. Die mit zahlreichen Mikrophotos und Strichzeichnungen belegte Abhandlung behandelt die Hauptlinien der Ascocarp-Entwicklung der typischen mit Algen vergesellschafteten Discomyceten einschließlich der Flechten mit apothecienförmigen Fruchtkörpern.

Das von REYNOLDS vorgelegte Buch bringt eine Fülle von Fakten, teils Interpretationen früherer Befunde aber auch sehr viel Neues zur Ascomyceten-Cytologie und Fruchtkörpermorphologie. Die besondere Ausrichtung des Inhaltes der einzelnen Abschnitte auf die These LUTTRELLS wird für einige Teilgebiete nicht unwidersprochen bleiben, zeigt aber, wo für die Systematisierung der fädigen Ascomycetes weitere Arbeit geleistet werden muß. R. TRÖGER (Jena)

V. L. SATO and M. L. GEFTER (Editors), Cellular Immunology — Selected Readings and Critical Commentary. 647 S., 155 Abb., 211 Tab. London—Amsterdam—Don Mills, Ontario—Sydney—Tokyo 1981. Addison-Wesley Publishing Company. $ 21.50.

Die Verfasser haben nach Art einer Vorlesungsreihe solche Originalarbeiten anderer Autoren zusammengestellt, die entscheidende Erkenntnisse in der zellulären Immunologie vermitteln. Dabei wird von historischen und inhaltlichen Gesichtspunkten ausgegangen. In 10 Kapiteln werden die Kooperation der an der Immunantwort beteiligten Lymphocytentypen und -subtypen sowie Makrophagen, Fragen der Regulation der Antikörperbildung und Immuntoleranz, der Rolle des Thymus und der Molekularbiologie des Immunsystems behandelt. Den Arbeiten jedes Kapitels ist ein didaktisch gegliederter Kommentar vorangestellt. Dies ist eine originelle Idee, die nicht nur dem Lernenden den Einstieg in komplizierte immunologische Fragestellungen erleichtert, sondern auch dem mit der Immunologie Vertrauten interessante Aspekte eröffnet.

R. S. FRITSCH (Jena)

G. SCHIMMEL und W. VOGELL (Herausgeber), Methodensammlung der Elektronenmikroskopie, 9. Lieferung 1980, 10. Lieferung 1981. 50 S., 26 Abb., 108 S., 45 Abb., 5 Tab. Stuttgart 1980, 1981. Wissenschaftliche Verlagsgesellschaft MBH. DM 33,60; DM 73,60.

11 Jahre nach Erscheinen der ersten Methodenbeschreibungen sind nunmehr 2 weitere Lieferungen vorgelegt worden. Wie schon bei ihren Vorgängern werden aktuelle Thematiken aufgegriffen, ein Vorzug, der dem Loseblattsystem innewohnt und der sichert, daß die Sammlung modern und auf dem neuesten Stand gehalten werden kann. Von bisher 35 vorliegenden z. T. mehrfach bearbeiteten Kapiteln wird eines in Form von 5 Einzelbeiträgen erweitert.

Die vorliegenden Lieferungen behandeln die Thematik „Räumliche Rekonstruktion", ein Beitrag stellt dabei Vorzüge des Einsatzes der Holographie vor (Schlüsselnummern 4.4.2 bzw. 4.4.1): 1. D. HODER: Darstellung und Ausmessung dreidimensionaler Objekte mittels Stereo-Hologrammen, 2. J.-G. HELMCKE: Die Erzeugung von Stereobildern und Stereogrammen, 3. R. BURKHARDT: Die stereoskopische Ausmessung elektronenmikroskopischer Bildpaare und ihre Genauigkeit, 4. H.-J. ORTHMANN: Fehler bei der Tiefenbestimmung elektronenmikroskopischer Stereoaufnahmen, 5. J.G. HELMCKE: Darstellung der Ergebnisse der Stereoverfahren.

Die Beiträge überstreichen diese Thematik von der Anfertigung einwandfreier Stereoaufnahmen bis zur stereophotogrammatischen Auswertung des Objekts. Die erforderliche Theorie wird ebenso behandelt wie Hinweise zur praktischen Vorgehensweise gegeben und die eingesetzte Technik beschrieben werden. Die jetzt vorliegenden beiden Lieferungen sind ein wertvolles Hilfsmittel zur 3D-Rekonstruktion von Elektronen-Mikrogrammen sowie zur qualitativen und quantitativen Bewertung dieser Rekonstruktionen. E. UNGER (Jena)

H. Senger (Editor), Proceedings in Life Sciences — The Blue Light Syndrome. XVI + 665 S., 432 Abb., 85 Tab. Berlin-Heidelberg-New York 1980. Springer-Verlag. DM 98,00.

Der Band enthält die Beiträge der ,,Internationalen Konferenz der Wirkung von Blauem Licht auf Pflanzen und Mikroorganismen", die im Juli 1979 in Marburg stattfand. Die einzelnen Arbeiten sind zu folgenden Gruppen zusammengefaßt:
1. Blaulichtwirkungen auf verschiedenen biologischen Gebieten;
2. Photorezeptoren und primäre Reaktionen;
3. Genese von Carotenoiden;
4. Kohlenstoffmetabolismus und Atmung;
5. Wechselwirkung zwischen blauem Licht und Stickstoffmetabolismus;
6. Entwicklung von Chloroplasten,
7. Physiologie der Blaulichtwirkung.

Schon aus dieser Gliederung wird ersichlich, daß auf dem Symposium die verschiedenen Blaulichtwirkungen dargestellt und ihre Mechanismen sowie die möglichen Photorezeptoren behandelt wurden. Eine Betrachtung einzelner Beiträge ist in diesem Rahmen nicht möglich. Der mit dieser Problematik Arbeitende wird diese gut ausgestattete Zusammenstellung mit großem Gewinn nutzen können. Die Übersichtsarbeiten ermöglichen auch dem Fernerstehenden ein Einfühlen in dieses wichtige Forschungsgebiet. H.-E. Jacob (Jena)

M. P. Starr, H. Stolp, H. G. Trüper, R. Balows and H. G. Schlegel (Editors), The Procaryotes, a Handbook of Habitats, Isolation and Identification (in two Volumes). XXVII + 2282 S., davon 156 Seiten Autoren- und Sachregister, zahlreiche Abb. und Tab. im Text, 8 Farbtafeln. Berlin—Heidelberg—New York 1981. Springer Verlag. DM 880,00

Die ,,Prokaryotes" sind kein Lehr- oder Handbuch im üblichen Sinn sondern setzen Kenntnis der Biologie und Biochemie der Bakterien und Cyanobakterien voraus und stellen kritisch zusammen, was über Vorkommen und Lebensweise cytologisch, ökologisch, biochemisch oder taxonomisch gekennzeichneter Gruppen von Prokaryoten, kurz über ihr Habitat, bekannt ist, ferner über Methoden der Isolierung, Reinkultur und Identifizierung und wenden sich damit besonders an ökologisch interessierte Mikrobiologen. Neben den Herausgebern waren 177 Mitarbeiter als Verfasser der 169 in 23 Sektionen zusammengefaßten Beiträge beteiligt. Zwei Vorworte, — von R. S. Wolfe und von den Herausgebern, — betonen in einem kurzen Überblick über die Entwicklung der Mikrobiologie im XX. Jahrhundert die zunehmende Bedeutung ökologischer Gesichtspunkte und den ,,organism approach" (R. S. Wolfe), der einen neu aufgefundenen Organismus als Ganzes zu seinem Habitat oder seiner ökologischen Nische und den dort herrschenden Lebensbedingungen in Beziehung setzt; — Entwicklungen, die sich schon Ende des XIX. Jahrhunderts andeuteten.

In der ersten Sektion geben 6 Essays, um den Leser auf das Ganze einzustimmen, eine allgemeine Charakteristik der Prokaryoten, indem sie auf die ökologische, cytologische und biochemische Mannigfaltigkeit und dementsprechend auf die Verschiedenartigkeit von Vorkommen und Lebensweise (einschließlich einer Menschen-, Tier- und Pflanzenpathogenität) hinweisen. Alle folgenden 22 Sektionen mit 163 Kapiteln demonstrieren mit einem Gemisch natürlich erscheinender und künstlicher auf Grund eines auffälligen Merkmals zusammengefaßter Gruppen das Chaos der gegenwärtigen Prokaryoten-Taxonomie. In Sektionen zusammengefaßt sind z. B. Spirochäten (61 S.), alle Actinomyceten (211 S.), Bakterien mit Gleitbewegung (109 S.), Scheidenbakterien und einige Bakterien mit Gasvakuolen (25 S.), Bakterien mit Zellwanddefekten (62 S.), alle phototrophen Bakterien einschließlich der Cyanobakterien (118 S.), obligat chemolithotrophe Bakterien (65 S.). Zusammengefaßt in einer Sektion sind auch die taxonomisch heterogenen Gruppen der Wasserstoff-Oxydierer, Methylotrophen, Kohlenwasserstoff-Oxydierer, Denitrifizierer, der dissimilatorisch Sulfat- und Schwefel-reduzierenden Bakterien, der Methanogenen, zusammen mit Halobakterien und mit den extrem thermophilen Gattungen *Thermus* und *Sulfolobus* (132 S.). Dem gleichen Prinzip einer teils natürlichen, teils noch künstlichen Gruppenbildung folgt auch Bergey's Manual 1974. — Die farbigen Tafeln zeigen Fruchtkörper von Myxobakterien, Zellformen und Standorte von Eisenbakterien, die Abbildungen im Text typische Wuchsformen und cytologische Merkmale. Im Text hervorgehoben und leicht auffindbar sind Färbemethoden, Verfahren der Isolierung und Kultur, Zusammensetzung von Nährsubstraten. Literatur ist am Ende der Kapitel zusammengestellt.

Die ,,Procaryotes" sind ein ausgezeichnetes Hilfsmittel bei experimentell-ökologischen Arbeiten und bei Verstößen ins Unbekannte. Nur einen etwas handlicheren Habitus bei geringerem Gewicht hätte man sich für den praktischen Gebrauch gewünscht. W. Schwartz (Braunschweig)

P. A. TRUDINGER, M. R. WALTER and B. J. RALPH (Editors), Biogeochemistry of Ancient and Modern Environments. IX + 723 S., 264 Abb., 149 Tab. Berlin—Heidelberg—New York 1980. Springer-Verlag. DM 88,00.

Das vorliegende Buch ist der Tagungsband des 4. Internationalen Symposiums über „Environmental Biogeochemistry" (Canberra, Herbst 1979). Der Titel sollte nicht darüber hinwegtäuschen, daß „Environmental" oder der Bezug zur Umwelt eigentlich schon in der Silbe -geo- enthalten ist. Läßt man die Themen des Symposiums für den Umfang dieses interdisziplinären Forschungsgebietes sprechen, so reicht es nach diesem Buch von der Entstehung des Lebens, seiner Entwicklung und seiner Äußerungen in geologischen Zeiträumen bis zu den heute ablaufenden geologischen und natürlichen oder vom Menschen verursachten umweltchemischen Prozessen sowie, anwendungsorientiert, den Möglichkeiten des Erzaufschlusses durch Mikroorganismen und den Wirkungen des Bergbaus auf die Umwelt. Die einzelnen Beiträge sind ganz unterschiedlich in Qualität und Umfang und repräsentieren einige der genannten Gebiete nur symbolisch. — Der erste, schwach besetzte aber gehaltvolle Abschnitt des Buches über biogeochemische Wechselwirkungen in geologischen Zeiträumen enthält neben dem Einführungsvortrag von P. CLOUD 8 Beiträge, u. a. über die Chemie präbiotischer, extraterrestrischer, biologischer rezenter und fossiler Porphyrine, über Photosynthese, Sauerstoff- und Ozonentwicklung in der Erdgeschichte, zur Aussagekraft gebänderter Eisenablagerungen über paläogeochemische Bedingungen und Spekulationen über die Genetik früher Lebewesen. Der größte Abschnitt über rezente biogeochemische Prozesse (38 Beiträge) beginnt mit einer locker und recht summarisch geschriebenen Erläuterung und Abgrenzung des Symposiumstitels durch T. D. BROCK, aus dem hervorgeht, welch universell gebildeter Naturwissenschaftler der Biogeochemiker als Spezialist doch sein muß. Die Biogeochemie als Arbeitsgebiet deckt sich in ihren rezenten Aspekten offensichtlich hinreichend genau mit einer offen verstandenen physiologisch ausgerichteten Ökologie. Die Originalbeiträge dieses Abschnitts umfassen ein recht buntes Muster von Fallstudien und Fallbeschreibungen, die von den Bakterienpopulationen antarktischer Seen über Sedimentation und Kohlenstoff-Flux in Seen, Methanbildung in Sedimenten, Beiträgen zum Schwefel-, Stickstoff- und Kohlenstoffkreislauf und deren Beeinflussung durch den Menschen, mikrobielle Eisen- und Manganoxidation bis hin zur Analyse von Pesticiden in Umweltproben reichen. Hervorgehoben seien ein methodischer Beitrag über Bestimmung von Bakterienmasse in Sedimenten, 3 Arbeiten zum Kohlenwasserstoffgehalt von Algenmatten und Sedimenten unterschiedlichen Alters und eine Arbeit über die Beeinflussung des Ozongehalts der Stratosphäre durch den Menschen. Grundlagen und Neuentwicklungen zu mikrobiellem Aufschluß und mikrobieller Laugung von Erzen kamen auf einer dem Symposium angeschlossenen Laugungskonferenz zur Sprache. Die Mehrzahl der 31 Beiträge befaßt sich mit Aktivitäten von *Thiobacillus*, u. a. in einem Vergleich von mikrobieller und chemischer Laugung. Andere Arbeiten beziehen sich auf Uran-, Antimon-, Wismut- und Manganerze oder technische Aspekte. Mehrere Beiträge sind der Beseitigung von Bergbaufolgen von der Abwasserbehandlung bis zur Begrünung von Abraumhalden gewidmet. Das drucktechnisch gut ausgestattete Buch wird bei den Spezialisten der genannten Themenkreise die ihm gebührende Anerkennung und Wertschätzung erfahren. Bedauerlich ist, daß sich infolge der eigentlich zu breiten Streuung der Themen nur ein eingeschränkter Nutzen des Buches für den individuellen Interessenten ergibt. A. C. SCHWARTZ (Bonn)

S. ZOTTER, CHRISTA KEMMER and M. MÜLLER, Mammary Tumor Virus, Investigations in Mouse and Man. 164 S., 29 Abb., 8 Tab. Jena 1891. VEB Gustav Fischer Verlag Jena, Elsevier/North—Holland, Scientific Publishers, LTD. M 69,00.

Diese Monographie ist einem ja schon klassischen Tumorvirus gewidmet, und eine große Menge an Primärliteratur wurde aufgearbeitet und dem Leser im Zusammenhang nutzbar gemacht. Leider kommt aber die Darstellung der Molekularbiologie zu kurz, wodurch sich die Autoren den Zugang zur Erklärung der Eigenschaften des Erregers und seiner Wechselbeziehungen mit dem Wirt erschweren. Einige Darlegungen auf molekularbiologischem Gebiet sind außerdem nicht völlig korrekt. Beispiele dafür sind solche Formulierungen wie „. . . to distinguish . . . between endogenous (chromosomal) and exogenous (extrachromosomal) transmission (S. 56), „In contrast to somatic proviral DNA, germinal proviruses are present in all body cells" (S. 62), „. . . endogenous viruses may also be transmitted extrachromosomally" (S. 63) oder in Abb. 15 (S. 61) die Gegenüberstellung eines „somatic provirus" und eines „genetic provirus". All dies ist nicht direkt falsch (wenn man weiß, was gemeint ist), aber mehr Klarheit des Ausdrucks scheint angebracht.

Vielleicht kann man vor allem das molekularbiologische Niveau in der 2. Auflage verbessern, die dem Buch trotz dieser Kritik zu wünschen ist, da es viele wichtige Erkenntnisse über dieses mäusepathogene Virus und seine Varianten sowie interessante Beziehungen zur möglichen Virusätiologie menschlicher Mamma-Tumoren zusammenfassend darstellt. D. H. KRÜGER (Berlin)

Bei Fragen zur Produktsicherheit wenden Sie sich bitte an:
If you have any questions regarding product safety,
please contact:

Walter de Gruyter GmbH
Genthiner Straße 13
10785 Berlin
productsafety@degruyterbrill.com